SIFTINGS

American Land Classics

CHARLES E. LITTLE, SERIES EDITOR

American Land Classics is a quality paper-
back reprint series that makes available
again books with the stature of "classics" in
the literature of landscape, nature, and the
American "place." Each book is a facsimile
reprint of the original edition and is hand-
somely produced on acid-free paper for
your permanent library.

SIFTINGS

BY

JENS JENSEN

Foreword by Charles E. Little

Afterword by Darrel G. Morrison

THE JOHNS HOPKINS UNIVERSITY PRESS

BALTIMORE AND LONDON

Preface and afterword to the Johns Hopkins Paperbacks edition
© 1990 The Johns Hopkins University Press
Printed in the United States of America on acid-free paper
00 99 98 97 96 95   5 4 3

Johns Hopkins Paperbacks edition, 1990

The Johns Hopkins University Press
2715 North Charles Street, Baltimore, Maryland 21218-4319
The Johns Hopkins Press, Ltd., London

Originally published in a hardcover edition in 1939 by Ralph Fletcher
Seymour, Publisher, Chicago, Illinois, and reprinted in a hardcover
edition in 1956 by Ralph Fletcher Seymour under the title *Siftings, the
Major Portion of "The Clearing," and Collected Writings.*

Cover illustration and frontispiece: The two photographs were taken by
Charles F. Davis at The Clearing in 1988, and they show yellow lady's
slippers in bloom and the entrance drive to the Lodge.

LIBRARY OF CONGRESS CATALOGING-IN-PUBLICATION DATA
Jensen, Jens, 1860–1951.
Siftings / by Jens Jensen. — Johns Hopkins paperbacks ed.
p.   cm. — (American land classics)
ISBN 0–8018–4021–X (alk. paper)
1. Landscape architecture.   2. Landscape.   3. Gardens—Design.
I. Title.   II. Series.
SB472.4.J46   1990
712—dc20   89-43556 CIP

A catalog record for this book is available from the British Library.

This book is dedicated to the memory of my wife, who was a true fellow-passenger on the ship of life.

CONTENTS

FOREWORD

Jens Jensen and the Soul of the Native Landscape

WHAT YOU are holding in your hand, this brief memoir published in 1939 by a Danish-born landscape architect named Jens Jensen, is a treasure—a treasure of a special sort that will delight readers and collectors who love those rare, out-of-the-way, "secret" books that are unsung but really quite remarkable. Over the generations, *Siftings* has been one of these books, an out-of-print volume held dear by a small group of insiders—in this case, landscape architects and garden designers with a literary bent—who have laboriously photocopied its pages for their friends and colleagues, or, failing that, grudgingly loaned the whole book with dire threats of bodily harm should the recipient not return the treasure promptly.

The value these Jensen fans (and they are just that) have attached to the work is not misplaced. *Siftings* is, to be sure, compelling reading for those interested in the creative design concepts of a man whom the *New York Times* described as the "dean of landscape architecture" on the occasion of his death, at age 91, in 1951. But it is also a distinguished literary work on our relation to the natural world and on the conservation and use of the land—particularly in the Middle West, which was Jensen's adopted home, his great love, and the source of his literary inspiration.

Jensen believed in the "soul of the plains"—the great prairie rivers, the hills with their "storm-beaten" pines, the gentle valley meadows, the lakes that provide mirrors for the reflections of "the poet and the dreamer," the old oaks "with roots deep in native soil." And these became the materials for his landscape parks and gardens, from the city parks of Chicago to Henry Ford's great estate, Fair Lane, in Dearborn, Michigan. The soul of the native landscape was so compelling to Jensen, in fact, that after retirement from his design practice at 75 he established "The Clearing," a "school of the soil," he called it, to which students, mainly in landscape architecture, would come to study not so much the techniques of the trade as the relationship of man and land while they engaged in hard physical labor in the fields and woods.

Born on a farm near the village of Dybbol, Denmark, in 1860, Jens Jensen emigrated to America in 1884 after graduating from agricultural college and serving a compulsory three-year stint with the German army, which occupied his part of Denmark at the time. According to Darrel G. Morrison, dean of the School of Environmental Design at the University of Georgia and a long-time admirer of Jensen, the army experience (most of it spent in Berlin) led Jensen to an intense dislike of any kind of authoritarianism or institutional constraint. He became a true populist democrat and was ready for America, especially the Middle West, where the strain of populism ran deep. Though from a prosperous family, the venturous 24-year-old Dane started at the bottom when he arrived in the New World—first working as a day laborer in Florida and Iowa, then finally landing a permanent job in Chicago as a gardener for the West Chicago Park District. In time, he rose to become superintendent and chief landscape architect. Then, in 1920 he resigned to establish an independent design firm in Ravinia, Illinois.

Though his beginnings were humble, Jensen was to become, in retrospect, one of the most influential designers in the country and the leading exponent of the "Prairie School" in landscape architecture, which, like its better-known analogue in the architecture of buildings, especially those of Louis Sullivan and Frank Lloyd Wright, featured natural materials and forms and stressed the fitness of the designs to the indigenous qualities of a region's landscape. Though the Prairie School approach languished after the 1920s, Jensen's view that we should make our designs harmonious with nature and its ecological processes was to become the preeminent theme of modern American landscape architectural practice.

But what of Jensen the writer? Here he might well remind you not of other design geniuses, like his friend Frank Lloyd Wright, but of Thoreau and Wordsworth in his extremely romantic (some might say sentimental) insistence on the personal apprehension of landscape and nature. Like Thoreau, Jensen would "front" the essential facts of nature and draw unexpected truths from them: "The study of curves is the study of life itself. Curves represent the unchained mind full of mystery and beauty. Straight lines belong to the militant thought" (*Siftings*, p. 34).

And it is almost impossible not to be reminded of Wordsworth in Jensen's description of sitting, "in deep reverence," of an evening on some ancient Swedish burial mounds overlooking the northern moors. Jensen writes: "Then, as night crept over the heath, the lone song of the heath bird would testify to the seriousness of the land and to its glory" (p. 19). Wordsworth is evoked, too, in matters of the aesthetics of landscape as applied to garden design, for this was the poet's lifelong interest and semi-vocation as the designer of his own gardens and those of others.

And so, remarkably, the very modern-seeming Jensen was as much a man of the nineteenth century as of the twentieth, a man who derived from the days in which the laying out of grounds was viewed as no less an art than poetry and painting. He was only a generation away from Thoreau, who died so prematurely in 1862. And he missed overlapping Wordsworth by only ten years. And yet, throughout the graceful observations and images of *Siftings*, much useful information is imparted for the contemporary student of landscape and garden design.

In the end, of course, Jens Jensen was a lover of the land, a man whose life and work were as inseparable as they were joyous and constructive. In this regard, his little book becomes a monument, an evocation of the "soul of the native landscape," which is the proper study of us all.

CHARLES E. LITTLE

PREFACE

BY THE open fire memories of earlier days came back for rehearsal, memories of my wanderings in many lands, memories of my own little world in the heart of Mid-America.

For more than thirty years notes had been recorded without any thought of putting them into readable shape. But here they are—experiences with the soil.

ART HAS ITS ROOTS
IN THE SOIL

TRAILS are the footprints of the ages. They mark the history of man. They were the course of early communication between races and countries. One of these trails, like a silvery thread winding its way over precipitous cliffs, still leads to my pioneer log house. The Indian, on hunting expeditions or in conflict with other tribes, trod this trail, because from here he could see afar. Pioneers followed him for adventure and for gain. But it was love of the primitive that led us here. It is a storm-beaten trail, where the elements roar and shriek in their fury.

On an outpost, challenging the west wind, a little colony of princess pine has found a home along this trail. It is a friendly group, unassuming in dresses of delicate pink, its members embracing each other in love. Real princesses they are. Theirs is a message of profound depth. They sing of love and of nobility. They are pioneers that dare face the unknown because *they* are a part of it. They give a sermon never to be forgotten. As we wind our way down the trail, we hear their song above the roar of the turbulent sea.

The world of the little princesses is a world of truth and friendship. It vibrates with the song of past ages, of infinite youth and freedom, and man has never changed nor added anything to its sermon of beauty.

1

Siftings

It is in the understanding of this world—a world not of our making—that life becomes richer. Great men and women have worshipped at its shrine, and therein lies their greatness. From this world each receives riches in the measure that he is capable of understanding the message of the infinite as the Master reveals it in the many sermons of the primitive.

A people might live amongst beautiful surroundings and fail to understand the message of these surroundings. Yet, who can say, but that in the event of time a leader might grow out of such environment, as the thought of beauty developed from generation to generation in the soul of the people who lived there, until it came to fulfillment in some form or another, just as the great tree of the forest is an expression of all the trees surrounding it, its friends and companions that contributed their share towards its growth?

Knowledge and understanding of the out-of-doors reveal to one's mind motives and forms. These motives and forms are nothing to be copied, nothing to imitate, but they serve as an inspiration to sleeping forces that eventually will bear wholesome fruit. Art grows out of native soil and enriches life as a people attempts to express and develop this growth. It is contemporary to life itself and is fastened in the chain of human endeavor. It comes from within, stimulated by environments and influenced by the customs and habits of a people.

2

Art Has Its Roots in the Soil

The art of living is the accomplishment and the fulfillment of our task during our sojourn on this earth. Those who observe the rules of life are a part of the endless chain of evolution and are happier human beings than those who have drifted into a self-centered life of excess, of arrogance and conceit, destroying themselves and leading others on false trails.

It is quite essential to understand the soul of our own environment and of our own country before we can appreciate and understand the arts and intellectual efforts of other people and the forces that lie behind them. To know one's mother tongue well is more important for great accomplishments than the knowledge of many foreign languages. To admire the work of one's own people, to be proud of the achievements of those who excel in the part they play in one's own community, and also to understand the work of others, brings us nearer to an ideal life for all.

Quite naturally, since I have spent almost a lifetime in the field of landscaping, my understanding of this art is greater than of all others. Landscaping is a composition of life that unfolds a mysterious beauty from time to time until mature age. All other arts are founded on dead materials. In these materials there is no growth. The thought they express may grow, but there is not the freedom, nor the mystery of the infinite, to as great a degree as in landscaping. Compare a growing tree with a monument of stone or mortar, which is definitely shaped, never to change. The tree's whole structure and its promise for the tomorrow are not surpassed on this earth. In passing, the fallen giant soon develops new beauty by feeding new growth, which extends its life into the far off future.

It is often stated, "the art of making landscapes is just a branch of architecture." What comparison is there between the creating of a building, which fits into a narrow and limited space, and the creating of large pastoral meadows where the horizon is the boundary, ever changing in light and shadow with the clouds above, with the light of early morn, at eve when the rays of the setting sun cast their reflection upon the earth, in the silvery moonlight, and in the changing colors of spring and summer and fall and winter? Such are the keys to landscaping.

The landscaper must be imbued with an imaginative mind. If his work is that of a master, it retains youth and vigor into an indefinite time unfathomed by man. The landscaper must see the tree in its full beauty hundreds of years hence. A painting or a cathedral is limited in this respect. Decay starts at its completion. Think of the giant redwoods of California and their forefathers, which probably date back long before the history of mankind! Each tree, each shrub, each flower expresses an individual beauty fitting for a certain landscape.

A true expression of native talent is not found in the pompous gardens of large estates. For true expression you must look in the simple gardens of the common folk. Here is found a true art that has grown out of the soil and out of the heart of those people. They belong! They fit! They tell the true story of the loving hands which created them.

It is wrong to think that landscaping is a collection of specimens from all parts of the world. Museums are not made to live with. The finding of plant varieties is a scientific venture, fine and noble in itself, but it must not be confused with art, as is so often done. To be inspired by and to create parks and gardens out of the beauty and composition of our native landscape is a much higher accomplishment than to form a garden with varieties of plants that have no intimate association with each other or with us and which at best become mere patch work influenced by the curious and scientific mind.

I have stated before that man is more or less influenced by traditions, but I have also stated that it is the spirit back of these traditions which is of value, which is a part of the race; yet it is the form which is usually copied. Art gone to decay emphasizes the form only, for the spirit has disappeared. The world would soon become uninteresting if all gardens and other works of art were copied and recopied. The understanding that other people in other places are developing what is within them, that they are showing their ability and their understanding in this higher sense of human relations, makes one's world rich. Our admiration for our fellowman in seeing him do his part well does not mean that we should steal or copy his endeavors. Such copying usually turns into cheap work, challenging the work of the master instead of enriching it. It is one thing to be inspired by the spirit of such work, another to copy its form.

To me the art of landscaping is more closely associated with music than with any other art. Its rhythm and its tonal qualities are as a folk song or a sonata. In its greatness it might be a symphony. For a friend, I planted a group of sumac on a hillock facing the setting sun. When autumn's frosty breath turned their leafy crowns into a flaming red, my friend called it the "Tannhauser Group."

Local color, the expression of the environments dear to us and of which we are a part, must be reflected in creative landscaping and be its motive. Through generations of evolution our native landscape becomes a part of us, and out of this we may form fitting compositions for our people. In this little world is found all that makes for a full life. Here we learn tolerance and charitableness, peace and friendliness. It is because of our lack of knowledge that we overlook this fact.

Any New Englander who has travelled westward often goes back in memory to his stone hedges, to the brook where many lovely days were spent in childhood, to the hills crowned with maples, and to happy days spent in these surroundings. A person born on the plains, having changed his home to the hills or mountains, longs for the vast open spaces, where the horizon seems to touch the earth and where freedom speaks louder than anywhere else in the world. Each type of landscape has a soul of its own.

I never feel the soul of the plains express their quiet and peaceful beauty in a more inspiring way than when I stand on one of the bluffs of a river, looking out over the vast prairies at sunset time with the purple horizon as the background. Peacefully the many homes, like small oases, lie scattered over the plains. Blue smoke from each oasis drifts heavenward, giving one a feeling of human life within at rest after a day's toil. Peace is over it all. This human picture is illuminated by the last rays of the setting sun far to the west, and it seems that God and man are rejoicing together at the end of the day for the peace and rest that are to follow.

But those who have seen the great prairie rivers in their turbulent mood during the flood season, tearing away everything in their path, cannot doubt that even on the plains there are elements that awaken the plainsman from his easy going way into great action. When the prairie winds howl in fury, uprooting trees and destroying homes in their path, then the plainsman is quite safe from falling into slumber.

Valleys have a message quite different from that of the plains. A little meadow covered with flowers, over which butterflies play and bees hum, is in itself a scene that becomes dearer and dearer as man grows older; and its power often becomes so great that it draws him back to his home of boyhood days once again to drink of the beauty of this meadow and to seek the little brook which carries his thoughts with its current toward the river, and from the river toward the sea, and across the sea to other continents.

Hills and mountains have romance. They belong to the daring. A group of storm-beaten pine trees at the edge of a cliff calls for adventure, but how ill these same pines fit on the level plains, and what a tragedy to see them here. Fir and pine forests have their message, which is quite different from that of the deciduous forest.

The old tree down the lane, the old oak tree touched by the storms and fires of many ages, with roots deep in native soil, has a message to tell. He has listened to the tramping feet of the Indians on their war path, to the pioneers in search of gain; he has listened to the cradle song of Indian squaws and to the cry of little children of the pioneers. He is now old, but beautiful in mature age. He speaks of the past, and he speaks of the tomorrow because his offspring will carry his memory into the distant future. In his old age he still sings the song of spring, life resurrected, jubilant and beautiful, with golden tassels in his hair silhouetted against the blue sky of his native land. Like a landmark he looms over the edge of the prairie, casting a radiant light on his environment. And when summer days are on the wane, great is his contribution to our autumn festival when all living growth joins together in all the colors of the setting sun for one last song before winter night calls for rest and slumber. To understand and appreciate the message of this old oak means more for a good life than all the books of man.

Inland lakes with their mirrors for reflection are for the poet and the dreamer, who here find food for thought.

Thus speaks the soul of our native landscape. Nothing can take its place. It is given to us when we are born, and with it we live. It speaks of freedom and friendliness. It speaks of a hope that gives joy and peace to the mind.

In the work of nature, man feels the Creator's mysterious power. Here he first recognizes his own limitations and his own significance. Here, there is nothing for him to boast of. It is not his alone; it belongs to all. In all its various moods, during all seasons of the year, it brings to us the message of the infinite.

Deep down in the primitive there lies the secret of the significance of life and of the infinite. It is a hidden, creative force visible only to those who seek and are attune with the Master's great creation. This force stimulates our thoughts, feeds our imagination, and urges us on to do our part well. Here the mind is free, free to follow its own inspiration. Trickery, cunning, hatred, jealousy, and the thought of doing injustice to others are all strange to such a world. Its whole and only thought is the fulfilling of the real and true demands of life. To be one with this creation, to understand its greatness and its purpose, its friendliness and its noble message, is to receive all that earthly life can ever give. Whatever human hands may render in real accomplishments, elevating the mind of mankind toward a higher goal, toward joy and happiness, the main source of inspiration will always be in the unadulterated, untouched work of the great Master.

The artist is a spiritual leader, and his message grows in importance as he comprehends the world in which he lives. He may lead his people into a wholesome, uplifting, and forward looking sphere. But all this is a growth stimulated by the message he receives from the creative force out of which he shapes his forms. A simple beetle finding its way through the tall grasses, the bee that hums from flower to flower, the wood thrush singing requiem at sunset, a colony of smiling flowers with their poetic charm, a sturdy tree in the winter landscape silhouetted against the sky and telling the story of many ages, a brook with a bit of rock protruding over its edge upon which a plant climbs in a daring way to receive a little sunlight that simmers through leafy boughs—each is a world by itself, full of mystery, charm, and beauty. Each is a book of great knowledge.

The little world about us has within it all the joy and happiness that we need. It has all the creative sources essential to making the man-made world beautiful for us to live with and enjoy. Each little world is somewhat different from the next one. Sometimes the difference is great, as between hills and valleys or between mountains and plains, but each speaks its own language. It is by listening to these messages and being inspired by these masterly creations that the world may become in its art expression as rich and as different as the expression of its local color. It is out of this, and out of this alone, that great arts can grow.

Understood by but a few, landscaping is still in its youth. It speaks of truth, and as the true significance of our life becomes better understood, it will grow and flourish. The landscaper belongs to the future. It is he who will weave the works of man and of the primitive into one harmonious whole, counteracting the scars made on mother earth through ignorance. He will oppose the enslaved thoughts of our machine age by singing with the freedom of our musicians and our poets of hills and valleys, of far-reaching plains, of intimate brooks, and of sea-going streams.

EARLY IMPRESSIONS

THE fjord stretched its arms out to the open sea—that mysterious world of sunsets toward which our forebears had set sail in long boats symbolic of the great sea worm which they believed encircled the world.

Great grandfather was a pioneer. He had dared leave the village where he had lived in contentment and security; he had dared build his home out on this point where the wind was not merciful and where the sea roared in defiance. Many are the stories told of the losses from the ravaging storms. I can still remember the time when people living along the sound had to seek a safe harbor in our farmstead. Shelter on this wind swept point was at a premium so great that grandfather surrounded the farmstead with trees—trees that he knew would grow and survive.

Time passed on, and wars swept over the land. The pioneer home was laid in ashes, and only a few trees remained to tell the story of his work. But like landmarks of the past these trees stood high over the farmstead, noticed by sailors long before their ship entered the fjord. In my boyhood days these trees became the gathering place of thousands of starlings in late summer, and I still can hear the grand chorus of these birds as they sang the sun down. From here they started on

13

their long flight to their southern home, many never to return.

My forefathers had grown up on this windy point. They had learned to love it; they had learned its secrets; they had struggled with the sea and the elements, and they passed their experiences on to those who followed them. Stories of the hardships and experiences of the two wars which passed over our land, and of the narrow escape of ships in distress, were always welcome topics when our family group gathered together during the long winter evenings. This understanding of the immediate environment, this knowledge of the mysteries of the sea and of the land across the fjord, and those stories of the vikings of an earlier period who dared the sea and the unknown for adventure stimulated the imagination and left their unmistakable mark on the growing mind.

When the first flowers appeared in spring, father made pilgrimages with his boys to the bluffs towering above the open sea. Can anyone realize what it meant to those who had been shut in for months to be greeted again by the warmth-bringing rays of the sun and the lovely green it brought forth, changing the earth into a new beauty? It was a delight to us boys to have father relate the story of these ancient bluffs and to find for us the first flowers of spring. Far out in the sea were markings where once the bluff had held forth against the ravishing sea that from year to year took its toll. This great love for the out-of-doors, for its history and

its beauty and its spiritual message, was so woven into
the lives of my people that it resulted in many festivals
during the year: Spring festivals, sunrise festivals, summer concerts in the forest, camp fires dating back to this
land's pagan days, and many public meetings were held
in the out-of-doors.

At the time of year when the woodruff filled the
forest with its sweet perfume and the lily-of-the-valley
tuned up the forest with its bells, we celebrated our
spring festival. At sunrise the orchestra joined with
the woodlands in a great spring symphony. This lovely
custom attuned one anew to the world about. Some
of these festivals had a tang of paganism in their
makeup—an outstanding one was St. Han's Night.
This night has since grown into a national event, when
the youth of the country gather around great pyres,
rejoicing in mirth and seriousness. Their song, "We
Want Peace In Our Land," resounds from fjord to
fjord.

School days brought more adventure. Three miles
we walked to school along roads lined with hedges full
of color and the song of birds in summer, and waist
deep with snow in winter. These hedges gave us children an education equal to that received in the little
village school, and to us far more interesting. For
myself, it helped lay the foundation for my life's work.
We used to spend our time hunting for rare and secret
plants, forgetting the hour for school and arriving
home late. This brought rebuke from both ends of the

line, but the time spent along these lanes still stands out vividly in my memory as hours of much romance and child adventure.

Father's fields were divided by untrimmed hedges. Hawthorn and blackthorn predominated as a protection against roaming stock. Like clouds of snow they appeared when in bloom in early spring. These hedges with their native life taught us to respect the rights of mute life. Here the birds found food and nesting places, and both the grain fields and we children profited by their association. Although the cultivation of the land dated back to the dawn of history, there was still an abundance of wild life. Rabbits and prairie chickens were abundant, and the hedge-hog gave us boys many thrilling hours. Late summer brought new adventure when blackberries and raspberries were in fruit. We were often thrilled by meeting the badger on moonlight nights feeding on the blackberries which grew so abundantly in the hedges. But the real delight came when the sweetbriar rose was in bloom. Its thorns were never merciful, and many a torn leg or arm was the result of too close contact with this lovely rose. My early love for it has never ceased, and it can be found far away from its native home, at the door of my log cabin.

It was quite customary to plant memorial trees in the school yard, usually oak trees, something that would last far into the centuries to commemorate great historical events. Great was my disappointment when forty-seven years later, on a visit to my homeland, I found the oak I had helped plant, together with my old school, removed, and a large square taking their place. Was this change for the better?

The beaten trail to school passed by a strange bit of landscape. The story was told that a castle had once occupied this space, but on account of the wicked life of its occupants it had slowly disappeared into the earth and had made the hole now filled with water. We children were most solemnly told by our elders that at midnight on Christmas Eve you could hear the bells ringing down below if your ears were trained for that sort of thing. Surrounding the pond was a bog in which grew plants not to be found elsewhere in our community. The calamus lived here, and we delighted in chewing the roots, just as children of today chew gum. When ice made the bog safe for exploration, we found cranberries and other plants strange to us. This little bog, where the footmarks of our ancestors were visible, became a shrine in our childhood minds.

Not far from the farmhouse was a small hill, a remnant of the work of the ice age. From that hill the eye could see the sound winding away like a great river toward the east, and beyond the sound appeared a belt of forest lands. Like the sea to the west, this purple horizon stirred our imaginations and created a desire for adventure. This hill became a lookout point from which we boys could view the world beyond. What a delight, therefore, when, as a committee to secure boughs to decorate the classroom in celebration of the birthday of the old schoolmaster, we started off over the fields toward what for many years had remained a mysterious world. Here we saw tremendous trees, the kings of beeches and of oaks. Deer scurried across the trail, and many other creatures strange to us. Singing we went, and singing we returned with our load of evergreen boughs.

It was a beautiful autumn day when father piloted me to one of our numerous folk high-schools, now cele-brated the world over. This school was situated out in the country, surrounded by lovely gardens, shady avenues, and broad lawns. I still love to dwell in the mem-ory of the many excursions into these environments for a better understanding of these lovely bits of land-scape.

The spoken word was the law of this school, and it still sounds in my ears as it did in those days now long ago. Here were born a love for home and fireside and for the brotherhood of man, and the intimate relation-ships among the students proved a real value for days that were to come.

Far off to the west, where water and land met, rose a chain of sand hills over which storms roared in defiance, piling up the sand in their fury. Here life was a struggle, a struggle with the sea and the sand. Beyond the shelter of these dunes stretched out the moors and heatherland of Jytland as far as the eye could see, level as our plains. It was a great world, where you felt the freedom and the will to be. On the edge of these moors were buried, back in the ninth century, a great King and a great Queen. Two burial mounds (some of the largest in all Scandinavia) still stand as mute testimony to this. On these mounds I would sit in deep reverence of the beauty before me—when the moors were covered with the purple of the heather and when at eventide church bells from far and near rang down the sun and a glow of orange and red told of the departure of the day. Then, as night crept over the heath, the lone song of the heath bird would testify to the seriousness of the land and to its glory.

From the shores of ancient Jytland—a country where every foot of land had its history, where the mind had received its first imprints of life, and where ideas had been molded into form—to the prairies of Mid-America, a new country, new to white men, where the Indians still roamed the plains and the buffalo enjoyed the freedom of vast prairies endless to the European mind, was like being torn up by the roots and transplanted into new environments with different habits and a different outlook on life.

There is, however, great similarity in the vastness expressed in the open expanse of the prairies and that of the sea. The sea has a distinct power of drawing one out, of arousing one's curiosity to investigate what is beyond the horizon. This without question had something to do with the urge of my sea-going forebears and with my own urge to explore the lands beyond the sea.

19

The prairies, too, have mystery. They, too, have a horizon that calls. Early treks across the great plains must have called the pioneer, urged him on to new adventure. When mist lies over the plains, they become a sea indeed. However, the prairies give a far more secure feeling than the sea. The prairies are inhabited; they are human. Like oases the homes of man are scattered over them. The sea gives anxious moments. In my boyhood home the anxiety for others was great when a storm was in its fury and ships were at its mercy. We never went to bed one night in stormy weather except with kind thoughts for those who were on the sea, hoping for their safe return.

Prairie folks are friendly. They have to be. They cannot hide behind cliffs or defy their enemies from a precipice. They are one family, so to speak. Only when fog lies over the land is the prairie man within himself. Then he may take soundings and see where he stands.

The prairies are still new in their development of character and in their influence on the creative mind, but that they are more powerful than the mountains can already be seen in the short time they have been inhabited by the white man. They are the Mid-American empire, the heart and moving force of our great country. To sit on the bluffs of the great Mississippi and let your thoughts flow with the stream is a great privilege, and to feel the vastness and the power of the great plains through which this Father of Waters flows is a greater privilege.

Man cannot comprehend the solitude and the spiritual message of the mighty mountains, but he can understand the message of that which he can measure in relation to himself, something not beyond his limited mentality. Great artists have already grown out of the prairies of our country, but many more are to come from this mighty force with its human touch and friendliness.

My early impressions by the sea are permanent. They cannot be discarded. But out on the plains they have had a chance to grow and develop into new forms out of which I planned my life work.

ENVIRONMENTAL
INFLUENCES

I BELIEVE that every community, be it city, town or village, has within it sufficient intellect for a fine culture fitting for that community. And when I settled in northern Wisconsin amongst farmer and fisher folks, the opportunity offered itself to test my theory. Not wanting to be different from my neighbors, I bought old abandoned log houses, built by the first settlers, and had them rebuilt by the neighboring farmers. In the village was a blacksmith who was a handy-man in many ways, as such village blacksmiths often are. He furnished the hardware, the candle sticks, the andirons, the chandeliers, and other little pieces of hardware that would be fitting for a log home.

Having succeeded in one adventure I tried another. A table and some chairs and chests for the linen were made by a farmer who was handy with his carpenter tools. I soon found another farmer who was handy with chisel and knife, and he carved the tables and chairs. Women weavers of the community supplied the home with rugs. In fact, they proved what I had earlier stated: that all talents could be found in all centers, fitting to their life. Dormant in the subconsciousness of these people lay the seed that needed awakening. With little encouragement their buried talents came to life. It was rural culture at its best— it was life.

Siftings

Because of the great variety of geological and climatic conditions of our country—the sunny and lazy south, the cold and rugged north, the great expanse of the plains and the desert, the Atlantic and Pacific coasts with their limitations—the opportunity for varied local individuality is unlimited and unsurpassed.

Through scientific developments we have today been brought closer together, and we seem to be dominated by the machine, which is trying to crush every bit of our God-given freedom to be ourselves. But the urge to be is much stronger and more powerful than all other forces put together. It will survive and express itself for the good of all mankind so long as man inhabits this earth, and his expression in the arts and crafts will be determined by the inspiration he receives from that section of the earth in which he lives and of which he is a part. In spite of all scientific knowledge and development, the arts that stand out nobly and beautifully, even in these days when the arts and crafts are struggling for freedom, have grown out of their own environments.

The world is divided into many little worlds, each of which exerts a certain influence on those living within it. This influence is found in the character of the people and expressed in their economic and spiritual life. Before modern transportation was discovered, these little worlds were more or less isolated, and the people's work, especially in the arts, had a better chance to develop in accordance with environmental influences and native intellect. Thus dialects, modes of dress, and variety in arts and crafts came about. Outside influences were rare and only brought in by adventurers or wandering troubadours who, in songs and stories, gave their impressions of foreign countries they had visited. Perhaps no section of the world, and especially of the old world, is more outstanding in that respect than the little island of Iceland with its renowned sagas, its beautiful wood carving, and its high intellect evenly scattered amongst every islander, poor or rich. Their knoweldge of the history of their own people and of the world far beyond them is outstanding in the life of peoples.

We are all aware of the Scotch character, which is much like the rocky and cold country the Scotchman inhabits. And the green and poetic island of Ireland, full of loveliness and charm, has just such characteristics as are today expressed by the humorous and witty Irishman. In the Scandinavian north there is a remarkable difference between the three races that unquestionably ages ago were one people, living, as some historians tell us, in the poetic lake country of Sweden. A change of environments has made a deep impression on the character of these peoples, on their art, and on their literature and their daily life. The influence of the Mediterranean, where the Latin races live, has created a distinct character quite different from that of the Germanic races.

In our country, although young in comparison with

Europe, the power of environmental influences can already be seen. For instance, two brothers migrate to America from Scotland. One settles in New England and the other one in the Carolinas. Their descendants can not be recognized as coming from the same stock. Environmental influences of the hot south have almost destroyed the strong and hardy characteristics of these northern people; whereas, the New Englander has maintained his Scotch characteristics.

The farther south a northern people migrate, the more degenerating are the influences of environment, due of course to climatic conditions which have changed their mode of living. Yet in the mountains of Virginia, Kentucky, and Tennessee the strong characteristics of a northern people have remained untouched, and although more illiterate than their northern brothers, their sound viewpoint of family life is as noble as can be found anywhere in this country. The mountainous influence has made them more daring than their neighbors on the lowlands. The people living on our coastal plains, where climatic conditions favor an easy-going life, are more musical, and we may expect, as time goes on and the cultural life of our people advances, great artistic accomplishments in song and music from the people of these regions.

In California, a much more recently settled country, environments seem already to have had an important influence on the lives of white men. The Californian is bound to the soil; he belongs; he is happy to belong;

his life is influenced by the forceful environments which are his state. No other section of America can portray such outdoor life as California. The Californian does not sit on his porch, as does his countryman across the mountains in the coastal plains. He is out and about the hills or on the desert. He learns to know his surroundings; the mountains and the canyons call him. He becomes himself, and in this lies great happiness. The fad of imitating others which exists today along the Pacific Coast is not due so much to the native Californian as to immigrants from the middle west and east. California will be California, and more so as days go by. No one can escape the power of this mountainous country with its striking local color.

Neither can the plainsman escape the power of the great plains of Mid-America, a power far greater than that of the mountains. The plains consist of many small worlds that may have local influences, but in the greatest measure the plains are one, and the expression of those who inhabit them will, in a general way, be the same. Some sections are more poetic, and as we look over those great poets of today who have grown out of the plains, we find that where the deciduous forests dominate, poets live and develop.

There is youth and strength and song in the rejuvenated green of deciduous woodlands as it bursts forth every spring in the warm sunshine. One delights in this rebirth of the woodlands after winter's cold. They vibrate with song and fill the soul with music and happiness. A colony of deciduous plants speaks of friendship at all seasons of the year. There is coolness in their leafy foliage in summer when the plains swelter in heat, and there is warmth in their brown and purple mantle on cold winter days when the northwester brings the icy wind from the far north.

The plains speak of freedom—earth and sky meet on the far horizon. There is nothing to intercept the vision from the infinite. Perhaps no section of America has so far shown as much power in the development of native art as the prairie country. Here lived and worked the great architect whose forms were inspired by the horizontal lines of the plains and whose decorative art grew out of his love for the native prairie thistle, a plant full of poetry and beauty. A pupil of this great architect has developed an architecture that has penetrated every civilized country in the world. And the architect who built the Capitol of Australia is a son of the prairies of Illinois. In spite of the hard-headed economists and the cold and greedy industrialists which the plains have produced, two renowned poets have grown out of the plains, and more are on the way.

Contrast these expressions of our rural country and those of the city with its prison-like homes, where the view is from wall to wall or down a straight, monotonous street walled in by stone and brick with nothing to feed the imagination or to inspire the emotions, where sunlight is at a premium and dark corners encourage vice and dishonesty, where mob psychology rules and men are led to copy others. In this entanglement of masonry the growth of cunningness and trickery, conceit, and jealousy and hatred is much greater than in the free and open country. I do not say that men and women of ability have not grown out of this barren pile of stone and mortar. Seeing the destructive influences of their environments they have tried to better them, but their success in combating the devastating influences these large centers produce has still to be proved.

Here fads thrive, good or bad. The average city citizen goes to sleep on the real importance of life. He becomes easy-going, easily satisfied so long as he can get an easy life. His spiritual senses are at low ebb, yet with his culture, which is more like a varnish that covers up many defects, he tries to dominate the country. The people of the city are easily swept into set rules and regulations, and the life of artificiality grows abundantly.

The city chap travels far beyond the boundaries of his home. When the city engineer lays out a new highway, he destroys the romance, the daring, and the excitement of strong contours. He is apt to make a toboggan-slide out of a concave curve by cutting too much at the top and filling too much in the middle. By so doing he ruins the contours of the country; he destroys the thrill of a concave grade. In other words, he interferes with the beauty of nature to a degree that has a softening influence on man's senses. Hazards are not always a detriment. More often they are a blessing. The man of the city generally lives on level streets, on artificial grades. He goes out into the country to find the reverse—to get the excitement of changing contours of the land, which he does not have at home. Then, to his own loss, he deliberately destroys them.

It is from the rural country, from the farming communities and the small towns and villages, that the real American culture will eventually come, as here it can freely grow. The city imports its culture from foreign sources and prides itself on so doing; it prides itself on imitating an older and now decaying culture. The city's own creation, the skyscraper, is already doomed as an unsound investment and as an uncalled for adventure in architecture.

Cities of the middle ages had a certain amount of mystery and charm, with their streets full of turns and sharp angles which exerted a certain influence on the character of the people, inspiring the development of the arts and crafts to a great degree. An amusing story is told about the visit of General McMahon, who, after his defeat in 1870, made a trip through France to ascertain the loyalty of his countrymen. In a certain French city with straight streets, no real enthusiasm was shown toward the general. Noticing this, the general smiled and said,

"How could there be any enthusiasm in a city of

straight lines, where there is no romance left, nothing to feed the imagination? How different are those people of the towns I visited where the streets are crooked and the corners well-defined, like cliffs in the mountains. There real enthusiasm was shown."

STUDIES

ONE bright March morning found me scraping the mud off a Chicago boulevard. At home in a closet lay a mat of drawings of some well known European gardens. The work I had been put to somewhat conflicted with these drawings, but a beginning had to be made, and bread and butter were essential. There was still a spark of my forefathers' desire for adventure, and this spring morning was to me an adventure.

The parks to which I was now attached, and with which I was to remain for a good many years, represented two definite ideas that had become a conflict within me, a conflict which was now raging—the natural or the architectural. Greatly did they disturb my peace of mind until I found myself. The great city of which I had become a part opposed my growing interest in the primitive landscape. But a curiosity as to the immediate environments of Chicago created frequent excursions, and I shall never forget my first discovery of what is now Jackson Park when its lowlands were purple with phlox. Where are these flowers today? Did the designers of Jackson Park forget them, or were they ignorant of them? During these early years of study in the Chicago parks I discovered the total absence of all the romance and poetry which was true

Illinois, and I asked myself why this great spiritual force had been overlooked.

Landscaping is just being born, and its birthright is the soul of the out-of-doors. The world is rich in landscapes in harmony with soil and climatic conditions. In the virgin forest you can read the story of creation. There is no repetition, no over and over again, but there are a multitude of ideas for the fertile mind to work with and shape into something that will inspire the race with a spiritual force for real accomplishments in the realm of art.

The study of curves is the study of life itself. Curves represent the unchained mind full of mystery and beauty. Straight lines belong to the militant thought. No mind can be free in a concept of limitations. Straight lines spell autocracy, of which most European gardens are an expression, and their course points to intellectual decay, which soon develops a prison from which the mind can never escape. The free thought that produces the free curve can never be strangled. If art means anything, it must be for the good, and the good can only be expressed in the will to be, both in form and spirit.

At the time my mind was searching for light, I had the good fortune to live for a short time in a European city where two major ideas of park planning had been developed side by side. One of these ideas was a French garden, executed on a large scale, with statuary and fountains and other ornamentations of a period that belonged to the time when the gardens of Versailles were created. These gardens were a good example of French garden architecture of a time when the ruling forces spent public money so lavishly that it continually brought into poverty the people who had to pay the upkeep. At the time I visited these gardens, their maintenance had to be put at such a low cost that they were actually in decay. Only on festive days was it possible to display the fountains in their full glory. These gardens were connected with a pompous boulevard which was used for army parades, exhibiting glorious military uniforms, or royalty parading under the long rows of stately trees—trees as pompous as the display that took place under their arching boughs.

Flanking either side of the boulevard were areas which were termed English parks—an idea that had come from another country and that was quite in discord with the French idea. Then one lovely summer day I was rather surprised, when wandering through the woodlands beyond these English parks, to meet a little garden enclosed within the deep shadow of the forest. This garden was architectural in its layout, but, nevertheless, it had a certain sense of freedom, more in keeping with the life of the people than the French copy. This little garden left an impression upon my mind that has never been entirely forgotten.

One Sunday afternoon I ventured beyond this little park and found myself on the edge of a meadow bordering a river. The meadow was full of sweet grasses and multitudes of flowers living their own lives and doing as they pleased under God's sunshine, giving joy to thousands. Upon following the path leading across the river I found a grove of friendly trees beside a natural spring. An enterprising individual had scattered a number of tables through this grove, where he served refreshments. Every Sunday afternoon during the summer months an orchestra filled the air of this restful nook with inspiring music. Soon this place became a weekly sojourn where we friends met to pass away an hour or two in friendly discussion and in listening to the music.

What a difference between the atmosphere of this hidden retreat and the sophisticated gardens on the other side of the river, even including the little English garden. Here you might come and go as you pleased, throw yourself on the grass, or sit down under overarching trees, centuries old, a remnant of a great forest that once covered this region. Many native flowers found a home here, and in a thorn thicket close by a thrush would sing at eventide, unharmed and unconcerned by the human gathering.

Here was most precious freedom. Beyond the open meadow were the distant horizon and the glorious sunsets. On the horizon one could see a group of hills. To my limited vision they were mountains. The urge for adventure and for climbing these hills became stronger and stronger as the summer passed by, and one beautiful autumn morning a group of us tramped ten miles to these distant hills. After we had climbed to the top, our efforts were crowned with a satisfaction that comes to all successful explorers. As memory takes me back, that summer stands out as one of the most important experiences in the forming of my professional life.

Untold motives and ideas are revealed to me in the out-of-doors, not to be copied, because man cannot copy nature, but from which to develop a folk song or a poem. The art of landscaping, like all other arts, is that of a fleeting thought that must be caught on the wing.

To produce mechanical and scientific effects in plant life is foreign to the true purpose of the landscaper and to the finer feelings of mankind. The clipping of plants into hedges and grotesque forms is mechanical, and the plant has been deprived of its freedom to grow into its full beauty. The skill of the landscaper lies in his ability to find the plant which needs not be maimed and distorted to fit the situation.

I remember a bit of limestone wall dancing in the moonlight amongst the deep shadows of trees, which gave me the feeling of a folk song. Suppose a clipped hedge had been put in its place. Would that have solved the same problem so beautifully and so well? Could the moonbeams have played as pleasingly on the clipped hedge as on the stone wall? And what a joy the stone wall was in the waning evening light, when the stone turned into a beautiful rose, or when it led the way in the darkness of night. Such a delicate expression has never been produced by shears.

There are those who think the parks in our cities should be more along straight lines to fit the straight lines of the city. It might be a great stunt to have our parks developed along straight and rigid lines, but how can the human form with its many curves fit into such a scheme? No town planner of great mind has yet proclaimed that straight streets in our cities make the city beautiful. What the city man needs is an expression of freedom in everything he comes in contact with to counterbalance the city's straight jacket, squared at all angles.

Parks and gardens of curves are always new, always revealing new thoughts and new interests in life. Straight lines are copied from the architect and do not belong to the landscaper. They have nothing to do with nature, of which landscaping is a part and out of which the art has grown. Landscaping must follow the lines of the free-growing tree with its thousands of curves. One might hope that in developing a beautiful outlook on life the youth of our country would learn the life of a tree and its tremendous importance.

OUR NATIVE LANDSCAPE

To CROSS the great plains of our country in a motor car, so that I might feel the expanse and the many types of landscapes of which we are a part, had long been my ambition. Great, therefore, was my enthusiasm when a group of us climbed into the shining new Ford and started on that long journey—long by motor car. I had traveled the same distance many times by train, but as the train traveled while we slept, the mind did not register the distance. When we reached Texas and spent our first night on the great plains, I thought of my departed friend's words when he spent his first winter in Greenland: "I was alone, alone on the icefield with the sky as my roof. It was a great moment, as it was there I found myself."

Much had I heard of the grandeur of the landscape of the northwest, Puget Sound and its environments, but when at last I had the opportunity of reaching that country, I was little thrilled with what I beheld. It lacked something I love—poetic charm and shy loveliness. Like all coniferous landscapes it was austere, dark, and gloomy compared with the deciduous forests. True, there were mountain meadows rich in flowers and majestic mountains, but I knew that here I could never build a garden, for the land sang a different song from the music and the rhythm I understood. The fresh and vigorous green of Mid-America was wanting.

Siftings

The land of the deciduous woods—the home of the poets—represents something fine and noble, with great vitality and strength. The sugar maple lives for many ages, though not so long as the white oak. A grove of these maples has within it the power of solemnity and beauty, and the oak and the maple are friends. They grow together, and they are tolerant of the smaller friends and associates that cling at their feet. The spirit of friendliness, as expressed in our hardwood forests, is not surpassed anywhere else.

I remember an autumn day in northern Wisconsin. We were descending a hill and had stopped for a last survey of a beautiful country before continuing the descent. Back of us lay a rugged landscape carved out by the ice masses of long ago. Birch, maple, oak, and beech covered these rugged hills, as they had done for untold ages. The sun had just gone below the horizon in the west, and the soft light of the afterglow reflected on the silvery lace-work of the beeches and the grays of the maples changed the landscape in a most fascinating way. Against the dusk of the approaching night the rays of the setting sun flamed high, illuminating the faint outline of the birches. Sky and land seemed to be in full accord. It was a mighty landscape, rugged, forceful, and yet in this light full of poetic rhythm.

It is often remarked, "native plants are coarse." How humiliating to hear an American speak so of plants with which the Great Master has decorated his land! To me no plant is more refined than that which belongs. There is no comparison between native plants and those imported from foreign shores which are, and shall always remain so, novelties. If, however, as is said, our native landscape is coarse, then as time goes by we, the American people, shall also become coarse because we shall be molded into our environments.

Every plant has its fitness and must be placed in its proper surroundings so as to bring out its full beauty. Therein lies the art of landscaping. When we first understand the character of the individual plant; when we enjoy its development from the time it breaks through the crust of mother earth, sending its first leaves heavenward, until it reaches maturity; when we are willing to give each plant a chance fully to develop its beauty, so as to give us all it possesses without any interference, then, and only then, shall we enjoy ideal landscapes made by man. And is not this the true spirit of democracy? Can a democrat cripple and misuse a plant for the sake of show and pretense? I am asking these questions because there are people on this earth who have used such methods in developing gardens that are admired by many. But these gardens exemplify none of the freedom which every democracy should possess.

Plants, like human beings, have their own individuality. Some plants to be at their best need association in a small colony or group; others love the company of multitudes, forming a carpet on the forest floor or in the open. Some speak much more forcefully alone, as,

41

for instance, the cottonwood with its gray branches stretching up into the heavens as a landmark on the plains. Some plants express their beauty in a lowland landscape and some on the rocky cliffs. Some fulfill their mission in the rolling hilly country, and some belong to the vast prairies of Mid-America. Others sing the song of sand dunes, still others of rocky lands. A grove of crab-apple trees on the edge of the open prairie landscape gives a distinct note to the plains. The timid violet sings its song and fits into a different composition from that of the robust aster.

To try to force plants to grow in soil or climate unfitted for them and against nature's methods will sooner or later spell ruin. Besides, such a method tends to make the world commonplace and to destroy the ability to unfold an interesting and beautiful landscape out of home environments. Life is made rich and the world beautiful by each section developing its own beauty. This encourages each race, each country, each state, and each county to bring out the best within its borders. Different minds interpret the message of each plant differently according to their understanding. Such messages collectively make the world rich in spiritual values.

Trees are much like human beings and enjoy each other's company. Only a few love to be alone. Some trees like to be placed in a small group, and with their branches interlaced they give an expression of friendship. Trees that like to stand out alone or in small groups become landmarks of their surroundings. There are trees that belong to low grounds and those that have adapted themselves to highlands. They always thrive best amid the conditions they have chosen for themselves through many years of selection and elimination. They tell us that they love to grow here, and only here will they speak in their fullest measure.

Where deciduous trees are close together in the forest, their trunks become tall and their leafy heads are far above the forest floor, permitting light to filter through so that life can grow and blossom beneath their sheltering heads. I have a deep love for the deciduous forests and woodlands. One wonders at the close relationship—let us call it friendship—of the lovely hepatica growing close to the roots of a century-old maple or oak, protected by the friendly arms of the sturdy giant which offers it shelter and warmth in the winter and cooling shade in the summer. One marvels at the unity and harmony of such companionship and the fitness of these plants in relationship to each other.

The Chinese say that a gardener must be a wise man because he must know the trees he uses and their growth for hundreds of years hence. How short a lifetime is to complete the study of the character and the beauty of the plants used in the composition of landscapes. Only last spring the real loveliness of our red maple was shown to me. A group of these trees were in bloom against the purple and gray background of the oak-covered sand hills across a marshy stretch. I had seen them singly scattered amongst oaks and sugar maples, but never had I seen them in groves forming a rosy mist over a dune landscape. These red maples, like pink clouds floating over the marshes and silhouetted against a background of gray hills, sang a song new to me that April morning.

My first impression of the illumination of sugar maples in the reflecting light of the after-glow of the setting sun came a good many years ago when I was driving over the hills bordering Lake Geneva, Wisconsin. The road had been planted with sugar maple many years before. These trees were in their autumn tint, and the afterglow was a fiery red. Its reflection in the tops of these trees produced a light that to me made the trees seem to be afire. Ever since that time I have always tried to place the sugar maple or the sumac in such a situation that the evening light would set their tops aflame.

One lovely autumn day in central New York we had reached the top of a hill just a little before the sun had set. Below us there stretched out toward the east a lovely valley of farmsteads and villages gay in the golden colors of the sugar maple. I was, thinking about the farmer women who had transplanted saplings from the woodlands to shelter and protect their new homes. I was thinking what a fine thing it was that in those days there were no nurseries and few railroads to transport things. These maples expressed the beauty of the hills of central New York. They expressed the local color of the region; they belonged to New York; they belonged to the soil, and they belonged to the homes of the people who settled there. This was one of the finest bits of rural landscaping I have seen in our country.

On a trip through Missouri one beautiful May day I saw the white oak with its golden tassels silhouetted against the blue sky of Missouri. I want to remember Missouri in that way. The white oak had been left on the steep hillsides and hill tops, and it greeted those who saw and understood the beauty of these hills. Soft gray branches against a wintry sky, rose and silver buds in May, a rich foliage in summer with a strong character—this is my vision of the white oak. A few years ago I had the pleasure of motoring through oak woods each morning, and I marvelled at their rich color.

Down in old New England the elm tree with its feeling of shelter and protection and its homelike expression seems to dominate in forming something that is typical New England. Evidently the pioneers considered the elm tree more fitting for their homes and village sites than any other tree. Perhaps here again it was the pioneer wife who planted the trees, as in central New York. In the elm there is a real domestic feeling that perhaps no other tree possesses. Out here on the plains it is associated with the level and the moist country. You will find it in bloom mixed with the blossoms of the yellow birch in the bottom lands of the northern part of the Mississippi River. The different shades of bronze of the elm heads and the yellow birch in full bloom against the purple bluffs of the Mississippi portray something monumental, but also delicate.

And on the plains the cottonwood is a sturdy chap. He dares it alone. The storms seem to have no effect on his many-branched head. But plant him where moisture is scarce and his freedom curtailed, and watch what happens. His arms become torn and broken by the wind. The reddish-brown catkins of the male cottonwood swaying in the spring breeze is an unforgettable sight. And what a delight on a cloudy autumn or winter day when the sun breaks through the clouds and its soft rays strike the cottonwood's silvery grey head! It then illuminates the surrounding country as if it were there to give life to an otherwise dark winter landscape.

One cottonwood, a landmark that could be seen by sailors from afar out at sea, measured over twelve feet in diameter. Through jealousy it was destroyed, and when its life's history was studied, it was found to be more than six hundred years old. For many years the lower part of the trunk had been hollow. The height of the hollow was nine feet, and in it a man could sit on horseback. Evidently the tree survived without a tree surgeon, as the hole must have been more than two or three hundred years old. Perhaps a prairie fire caused the beginning of the trouble in the tree's early youth. When it was killed by a jealous woman to get revenge on her neighbors, there was not a dead branch on the stately tree, and the first branch was more than seventy-five feet above the earth.

When you see the white birch you know you are in the north. The canoe birch is really the poet of our northern woods. A pure stand of birch trees is quite a revelation, especially to those who are not used to such a picture. Birches and moonlight are real companions. No other tree speaks so beautifully as the birch in full moonlight. The reflected light of moonbeams playing on the white bark of the birches illuminates the woodlands with a surprising clearness. A group of birch trees placed outside a window, so that they receive the full light of the moon, will lighten up a room in a most startling manner.

On a hill in southern Wisconsin stands a little group of birch trees which we visited for many consecutive years at the time the shad was in bloom. To see the fleeting white blossoms of the shad entwined with the white bark of the birch trees was a sermon we were willing to go more than one hundred miles to listen to, and we always felt we were more than repaid for the trouble.

The north brings us not only the birch but also the aspen. In groups upon the woodland border it gives the landscape a decided musical note. One can almost see the woodland nymphs dancing to the notes of the violin. There is always a feeling of spring in the aspen, and with the gray dogwood in the lowlands they sing of spring when winter is still in the air.

As the oak is the dominant tree of Illinois, so the beech represents the remnants of what once was a great forest over a large part of Indiana. A beech forest is a noble thing, with its straight silvery-gray trunks terminating in a head of fine lacy branches. The beech, too, seems to be tolerant of its neighbors and smaller friends, as shown by the rich flora that grow on the floor of our beech woods. And yet the beech loves its own company pretty well. What a loss it will be to the landscape of Indiana, where the beech reigns supreme, when the last beech woodlot has been destroyed. I wonder if the Hoosier State will then create as many poets as it has heretofore?

Down in central Illinois the honey locust is at home, and in some sections is known as "The Farmer Wife's Tree." This name has been given it because of the fact that here again it was the farmer's wife who went into the wooded areas along the prairie rivers for the locust saplings. She felt more of a kinship for the locust than for other trees. There is a certain refinement about this tree, and in its golden-yellow autumn color it gives a soft light to the landscape.

Evergreens are at home in rocky or sandy lands. There they express something the deciduous plants can never do so well. White pine on a rocky cliff gives an expression of height and adventure, but white pine on the plains has nothing in common with the plains and nothing in common with their floral carpet. Many favor the evergreens in the plains winter landscape because they are green, not seeing the richness of the soft, delicate arms of the deciduous trees in harmony with their winter surroundings. The red cedar, which turns into a lovely brown during winter and blends beautifully into the browns and grays of our deciduous winter woodlands, is different. It belongs with these woodlands, just as the dwarf juniper belongs on our prairie river bluffs. Have you noticed its feathery aspect and changing colors in winter? Contrast this softness with the almost brutal expression of many of the imported varieties. In our cold material life we need a touch of the tender native juniper.

To me it is stupid to transplant trees into an environment they dislike and in which their length of life will be shortened and their real beauty never revealed. Of course the reason for this stupidity is ignorance and commercialism. It is a crime to deface a beautiful countryside, to mutilate what has been given us, by greed and ignorance. Landscaping will never flourish under those banners.

My path from the railroad station to a home I was landscaping passed a grove of red cedars. For a long time these cedars did not interest me. But one evening when I had been detained at dinner, I passed them in the full moonlight, and I was amazed by their deep shadows and their spiritual expression, which became mysterious and awe inspiring. Since then I have valued them in a reverential way. But they belong in groves, not singly like an exclamation point, arrogant, dominating their surroundings, singing their song alone without any association or any feeling for their neighbors.

Watch the purple ridges of deciduous woodlands from across Illinois prairies. Watch them when snow covers the fields. Get closer to them and feel the warmth streaming out from their warm texture to you. Then see man attempt to destroy this harmony, this noble character which is Illinois, by planting coniferous trees. What a travesty of so-called educated man's intellect!

Looking out of a train window one early spring morning, viewing the country as the train sped on and on, I was pleasantly surprised by what I later learned to be a western crab-apple tree in full bloom. It was our first meeting, and unforgettable. As time has passed, the affection I felt for the native crab-apple tree on this first introduction grew a hundred fold. There is nothing more profound on the edge of the woodlands in the prairie landscape than a grove of crab-apple—its lavender touch reflected on massive snow fields, turning into a delicate blue in the light of the day's afterglow. Or its pink blossoms in May bringing untold mystery and loveliness to the forest border where prairie and woodland meet. No more fitting tree can grace the gardens of our prairie dwellers.

A friend and I were coming up the Illinois River

one evening at sunset time. The afterglow was in all shades of yellow, and its full light fell upon a group of crab-apples a few hundred feet up the road, turning them into a delicate blue. How surprised I was to see them in this color, but my friend assured me they were crab-apples. From that day I have visioned this tree gracing the parks and gardens of my people. Neither you nor I can fathom the effect it shall have on our cultural life.

There had been no time for the preparation of an address I was to give at the University of Illinois, so on my journey to Urbana I hoped some worth-while thought would appear. Almost at my journey's end I observed the rosy head of a native crab-apple in full bloom. Sympathetic pioneers had saved it from destruction. They too had loved it. In its early life it perhaps stood on the edge of a forest where woods and prairie met. Since then the forest had been cleared away, but this tree was left in the open field. The storms of spring and fall and the icy blasts of many winters had shaken its tender body, even threatened its life; yet it sang the song of youth. More than one hundred summers had brightened its life. Like a bride in May it greeted us and other wayfarers who passed by. It, too, was a pioneer and was left to tell its story and to give of its beauty to the American of today and tomorrow. The message from this old crab-apple tree was what I gave to the students that day. I interpreted its lovely May song and its long life struggles.

51

I told of its fitness to the soil, of its pioneer courage and its will to survive and give of its beauty to its surroundings. That which fits, endures and records, not one lifetime, but many generations.

It is quite difficult to imagine a prairie landscape in Illinois without the hawthorn. Like a friend its outstretched branches, harmonious with the horizontal lines of the plains, greet you from afar. Its silvery lace-work against the purple ridges of the forest border add a note of poetry to our winter landscape. Like the crab-apple, it, too, is a pioneer. It wanders beyond the sheltering forest border and invites others to do likewise.

Hawthorns are always illuminating. Their horizontal branches are always outstanding. Some varieties carry their scarlet-red berries far into the winter, and when snow has covered the ground with a white mantle, they add a brilliant touch of color. Hawthorn branches thatched with snow under a canopy of a multitude of red berries surpass the winter charm of the much praised coniferous friends. Hawthorns used in the making of a landscape give breadth and bigness to the composition. The American landscaper will always find them in scale with the vastness of the country with which he works. I love them in the woodlands where they give mystery and a lacy note to their surroundings.

If I were taken to the state of Iowa blindfolded and there my blinds removed, I could tell where I was by the numerous native plums in hedge rows and in other places. I have never seen the plum more numerous anywhere. Its sweet-scented blossoms, changing from pure white to different shades of cream or pink, or even rose, are charming indeed. The native plum loves company and speaks most eloquently when in a group. In river bottoms its blossoms appear like clouds of snow drifting through the shady bottomlands. In the woodlands its blossoms appear like fleeting notes in a mighty symphony.

Who has not marveled at the delicacy of the flowers of the juneberry, or, as it is often called, the shad? To feel the real beauty of the juneberry is to see its frail blossoms intermixed with snowflakes on a stormy day in early spring—youth daring the tempest. The lesson of this plant is vigor and courage. It gives us hope for the tomorrow.

My first acquaintance with the redbud dates back a quarter of a century. It was on an excursion to the historic spot of Starved Rock on the Illinois River. Everywhere the bluffs were colored with the blossoms of the redbud. To me, who had never seen this plant before, it was a delightful experience.

Memory brings back a little incident in connection with this trip. Luncheon had been prepared for us in a little town nearby. The proprietor of the hotel had been kind enough to decorate our tables with carnations, but it was rather a weak greeting after seeing the blossoms of the redbud. How gay the dining room would have been if a little spray of this native plant that now was making the valleys of the Illinois River gay and festive had been placed on each table. Instead, we had a manufactured greenhouse flower. Man still seems far from understanding and appreciating the beauty of his native land.

Down in old Kentucky, amongst the limestone ravines and hidden brooks where the redbud and the flowering dogwood make the landscape gay in spring, a garden had been made. Recently it was my good fortune to pass through that section of the bluegrass region where this garden had been built many years ago. I was anxious to see the results of my labors, and to see what had happened during the years. Great changes had taken place. The former owner had departed. Following a cow trail we finally came to the old swimming pool in a sunlit glade in the woods that was now luxuriant with redbud and dogwood planted years ago. This pool was not just an outdoor bathtub, but a real swimming hole, attuned with the body of man and the woodlands surrounding it. My greatest satisfaction came when I saw the garden. It had found itself back in the bosom of old Kentucky. Man's hand had disappeared. Only his soul remained and, as it should be, in harmony with the hand of nature.

Swamps in the north are really the gayest of gay in early spring and late autumn. Here grow the small willows in various colors, together with sumac, dogwood, and huckleberry. Grasses and sedges in different shades of brown on the edges of these swamps become a tapestry of beautiful hues, warm and pleasing in the crisp air of spring and autumn. Man often tries to imitate this wealth of color by using exotic plants with differently colored branches, but he utterly fails in his purpose because he does not take into consideration that the coloring of these native plants is just for a short period. After that, they change into a somber green, quiet and restful for the warm summer days.

In central Indiana the prairie rose runs over farm fences, along the roadside hedges, and in the open glades of the woodlands. Wherever it grows it is a lovely bouquet, and its red berries over the snow in winter are as colorful as the rose in June.

"To do your bit well," is the motto of the witch-hazel. When all of its companions have lost their autumn glory and have settled down for their winter sleep, the witch-hazel, like a golden mist drifting through our woodlands, brings a new beauty to the late autumn. Have you seen the delicate blossoms of the witch-hazel in an atmosphere of purple? This pleasing shrub unfolds new beauty each day as its foliage disappears, until all that is left to sing the requiem are its lacy branches with their delicate autumn blossoms.

One cloudy April day, when threatening rain caused the west to be in a dramatic mood, we were scurrying along to reach some shelter before the worst might happen. A lone hazel bush, perhaps the last of a great colony, made us pause on our way. Why it had been spared I do not know, but there it was in a festive spring outfit. We were astounded by the attraction this simple plant possessed. The secret of it all was its yellow catkins against the threatening purple clouds in the west, bringing out their exquisite beauty. As they swayed in the breeze we could hear the cradle song of Indian squaws of long ago; we could hear the marching song of pioneers who had followed the trail toward the blue sea beyond the horizon. I had known the hazel since boyhood days, but I had lived almost an average lifetime before I saw its real significance and its charm. From that time on a hazel bush, backed by the purple branches of our native plum, has graced a corner of my garden, and every spring I wait for the spring song of its catkins.

Crossing some fields in central New York, I saw for the first time our native clematis (virgin's-bower) covering the fences. Never had I seen it grow so profusely before. With us it is rarely seen, except in shady nooks where man's foot does not tread. Like a timid child it hides its face from the madness of the world. I love to think of it as we find it here in shady nooks, when one sees the refinement of the white flowers.

When I first set foot on Illinois soil, the buffalo still roamed the western prairies, and the Indian still dared assume his rights against the white man. The primitive prairies of Illinois have not been entirely destroyed. Here and there has been left something of the primitive that the plow had not turned under. It seems a pity, rather a stupidity, that some section of this marvelous landscape has not been set aside for future generations to study and to love—a sea of flowers in all colors of the rainbow.

Along our railroad rights-of-way one meets the last stand of these prairie flowers. What a wealth we would have if our prairie roads could be lined with this rich carpet of colors, miles of flowers reflecting their colors in the sky above, or millions of sungods (sunflowers) in the strong prairie breeze nodding their heads to the sun that had given them their golden hue. But perhaps this is too much to hope for, as man seems unappreciative of these gifts.

One evening at the opera in Ravinia, an outdoor theater, when my thoughts were not much in sympathy with the foreign opera, as I did not understand the language, my eyes shot across the audience into an open glade in the woodlands where the brilliancy of the goldenrod in the path of the sun's afterglow gave to me an illumination worth a million operas, and I silently wished that those about me might have seen and felt this spiritual message from their native soil.

It goes without saying that many of these flowers which are so beautiful and fitting on the open prairies are often disappointing in a shady situation in the garden. Light and shadow and scale play a principal part in the art of making gardens. Some years ago I was honored with visitors from the University of Illinois. Amongst them was a famous photographer. He was looking for certain plants he wished to photograph, especially the tiger lily and the prairie phlox. The lily that once covered large areas in certain sections had practically disappeared. Only small groups or single plants were found. The phlox that once was so numerous on the wet prairies nearby had entirely disappeared.

On a cultivated meadow we found one plant. The professor disagreed with me that the plant he saw several hundred feet away was a phlox. I insisted on investigating, as I was sure I knew the plant well. We found it to be a phlox, a plant the plow had missed. The professor then determined that it was a new variety. I took my hat and placed it between the sun and the phlox, and the color changed. So the high priest learned a lesson that day. The lack of a knowledge of this difference has caused many mistakes in gardening.

The early creeping phlox often covers a sandy slope so densely with its white flowers that it looks from a distance like snow, but placed in a crowded border it would mean nothing. Its color value and its real beauty would be entirely lost.

It is not always the great outburst of color, like the blazing light of the setting sun, that touches us most deeply and brings into growth beautiful thoughts and ideas. The little dune violet, that hides away its sky-blue face from the gaze of man and the tramping feet of the many, speaks more profoundly and sings more deeply into the soul of man than the gay cactus plant with its large yellow roses covering the sunny slopes of the duneland within a stone's throw of the shy violet.

Never shall I forget a May day in the woods of northern Wisconsin when snowdrifts were everywhere. Along one of the snowdrifts, on a little sunny slope, the trailing arbutus was in full bloom. Years ago I had seen flowers alongside high snowdrifts in the Rockies and had marvelled at the close association of spring and winter. Now I had found the same lesson down on the plains of Mid-America. If you want to see the trailing arbutus at its best, where it sings of spring and sings of winter, you must see it in full bloom alongside a snowdrift.

I might go on indefinitely speaking about these many plants I have met throughout the years, each having a sermon for me. There are but few plants that do not love company. On the other hand, most of them are particular about their associates. The spiritual message or character of the individual plant is often enhanced by its association with other plants which are attune with it. Together they form a tonal quality expressed by an orchestra when certain instruments in chorus bring out a much higher and a much finer feeling than a combination of others. The different plants are then given a chance to speak their best.

I have often marvelled at the friendliness of certain plants for each other, which, through thousands of

years of selection, have lived in harmonious relations
But I have also found through years of experience many
plant combinations that are not friendly. Years ago I
experimented with a great many varieties of spring
bulbs. They were planted on a slope of our ravine,
which I thought was ideal for them. Some soon disappeared; others have remained for more than twenty
years but have not visibly increased. Others have run
all the way to the bottom of the ravine, and on their
march they have completely destroyed a stand of
maiden-hair ferns. That I made a mistake was evident.

When travelling through Maryland and other eastern states, I have been amazed to see the disturbances
caused by the introduction of the Japanese honey-
suckle, not only to the forest floor but to small trees
and shrubs. This shows the ultimate danger of trans-
planting plants to soil and climate foreign to their native habitat. The great destruction brought to our
country through foreign importations must prove
alarming to the future. Many of these importations
will in time become the sparrows of the plant world
and destructive to the beauty which is ours.

The motives and compositions in our native woodlands, our hills, our valleys, our river bluffs and our
swamps, our cold and rugged north, and our sunny
south are unlimited. Such wealth and such refinement
speak well for the art of landscaping in our country.
Examples have been given, but before a plant is used
in the composition, it must be tested—whether it likes

to be alone or in a group, whether it enjoys lowlands or highlands, whether its character and color sing with its surroundings.

My departed friend, our great western poet, Vachel Lindsay, had honored us with a visit. It was early morning when he called me to the open door where he was standing looking out over a clearing. There was a peculiar light over this little sun opening, caused by the reflection of the sunrise. The clearing was bordered by a simple composition of hardwoods with a few hawthorns, crab-apples, and gray dogwood scattered on the edge. The light had added an enchantment to this simple composition, and Lindsay, watching this, said to me, "Such poems as this I cannot write." Many years have gone by since then, many mornings and many evenings, and I have watched the clearing. I have seen it on cloudy days and in full sunlight, in the starry evenings and on dark nights and moonlight nights, but I have never seen it the same.

Light and shadow and their distribution during the entire circle of day and night are important fundamentals in the art of landscaping. Many of the fine and soft shadows in the color scheme of the Master's work are far beyond human hands to produce. But here is the real motive, the cue to the beginning.

COMPOSITIONS

WHO will admit that we are inferior to other peoples and have lost the courage and the vitality to be ourselves? Let me repeat what I have stated before: Art must come from within, and the only source from which the art of landscaping can come is our native landscape. It cannot be imported from foreign shores and be our own.

The suggestions nature offers are tremendous, and it is a great mistake to think that one mind can comprehend them all. One can love only a few things in life, and it is the things we love which take our whole interest and out of which we can bring forth the best we have.

It is far from the Atlantic to the Pacific and from the Canadian Border to the Gulf of Mexico, and within this great stretch of land many different landscapes appear. The soul of the great prairies is quite in contrast to that of the mountains. Inland lakes tell a different story from that of the sea with its vision of foreign shores. Woodlands do not sing with the same rhythm as peaceful meadows and babbling brooks. It is folly for the landscaper to think that one type of composition will fit them all. Each problem has a different answer.

As I look over the field of my endeavor, it seems to me that each locality had within it the possibility of certain definite expressions or motives. Fitting plants were needed to give the landscape a true note of harmony and beauty. Plants with characteristics fitting to that particular situation I repeated again and again throughout the plan. A native crab-apple, a group of violets, a bird bath with a cardinal flower, a virgin's-bower clinging to the crab-apple tree, and perhaps a dash of iris or phlox, or lilies if you please, make a garden full of song and poetry.

A few of the homes I have had the pleasure of planning give me pleasure to this day, although I would consider none the final word. They all seem to be a part of a long list of work, each one discarded when a new one came on the horizon to receive my full attention. Out of the many looms one place that has become dear to me during the years. It was a bit of oak and maple woodland, hundreds of years old. This woodland was protected on two sides by deep ravines. The struggle of some of the trees, scarred and scorched by forest fires, was very noticeable. Yet they survived to tell the story of pioneer days and of a time long before. Their roots were deep in native soil and their story far into the past.

In the heart of this woodland I cut a sun opening, or clearing, as I call it, and on the edge of this clearing I placed the home. Cutting into woodlands usually leaves some hard lines in the way of bare tree trunks. To overcome this, several hawthorns were brought in and placed at prominent points on the edge of the woodland. These trees through their natural spreading characteristics gave the clearing a feeling of breadth; they softened the cold cut I had made and gave the little sun opening a poetic feeling most essential to make it more than a hole in the woods. A few crab-apples and gray dogwoods that originally occupied some of the cleared space helped to soften the picture and give it a finished touch.

If you look from the window of the little home today, out across the open expanse into the background of oak and maple, you see a quiet pastoral scene of complete harmony. When the long shadows come over the land and a thrush sings of the glory of the day that is gone, when the west fills up with rose and lavender, the view across this bit of landscape into the ravine with its dark, mysterious depths, and beyond into the floating clouds in the evening sky, is one not easily forgotten.

There is a soothing sensation about this clearing. It is large enough to be in scale with the ancient oaks and maples surrounding it and small enough to give the whole an intimate note. What a delight to see the starry heavens through an opening of the leafy heads of the oaks, or the moon throwing its soft rays across one corner of the clearing, leaving the rest in deep mystery. At such times it becomes a sermon which awakens the best in the human soul.

Following the ridge of one of the ravines was an old Indian trail, and at the point where the two ravines met was a chipping station where Indians made arrowheads for hunt and for war. Along the trail and on the ravine slope dogtooth violets, trilliums, hepaticas, wild geraniums, and numerous other plants, all friends of the oak and the sugar maple, covered the forest floor. On a little elevation at the intersection of the two ravines was a fitting place for an outdoor stage—we called it "Player's Hill." Many plays have been given here, some fitting the moonlight and others the dark and stormy nights.

Just below "Player's Hill," on the slope of the ravine, the first council ring was built—a new adventure. In this friendly circle, around the fire, man becomes himself. Here there is no social caste. All are on the same level, looking each other in the face. A ring speaks of strength and friendship and is one of the great symbols of mankind. The fire in the center portrays the beginning of civilization, and it was around the fire our forefathers gathered when they first placed foot on this continent. This particular council fire is situated where it may be seen from up and down the ravine like a message of greeting to others. The smoke of the fire, illuminated by the moon, forms fantastic shapes which gently float over the deep and penetrating shadows of the ravine. Many of these rings have I built since this first attempt. When they are placed on school grounds or in playfields, I call them story rings. These rings are the beginning of a new social life in the gardens of the American of the tomorrow.

The ravines have been left undisturbed for nature

to work out her own problems. They give to us their depth, their mystery, their wild life, their secret nooks with rare plants. Their importance to the spiritual mind is much greater than if they had been improved by trails and bridges, something the floods of spring do not like.

The old Indian trail following the ridge became a fitting path to the garden, which was placed on the edge of the west ravine. Here it received the full afternoon sun and was hidden from the babble of the street.

At first, the making of this little garden was an economic matter. It consisted of small fruits and vegetables. In one corner was a place to rest, shaded by overhanging wild grape vines. But prosperity came along, and man became lazy, not caring to work in a vegetable garden; so a change was made. The vegetables and small fruits were forgotten, and the slope toward the ravine's edge was made almost level. This created a difference in grades at one end of the garden, and here a rocky ledge was formed. This ledge was symbolic of the rocky cliffs along the Illinois and Rock Rivers, which I so greatly admired when I first came to the prairies.

There is a remarkable nobility in rocks, weather-beaten and worn by water of past ages. Rocks, like trees, have a character all their own, and this character is emphasized when the rock is rightly placed. Usually only a few plants can be used in connection with these heralds of the past, a little moss and a few clinging plants. One should always keep in mind that the rock has a story to tell, and it should not be vulgarized by a conglomeration of unfitting plants.

This particular ledge was situated where it received a great amount of shade; so it soon became weathered, and green lichens and mosses made it quiet and restful. In the winter picture it became a delightful note in the garden with heads of a few goldenrods and asters silhouetted against its warm texture, and climbing over its ridge a few prairie roses gay in their winter fruit gave it a note of joy. I delight in visiting this nook in winter when snow partly covers the rocks and some rabbit has found shelter in a protected corner.

Below the ledge a pool was added to bring the play of fish and a friendly frog or two into the picture. And in front of the pool a refreshing green carpet on the level grade stretched away into the woodland border. A lilac bush that belonged to the early garden, which for sentimental reasons could not be forgotten, found a spot near one of the garden's resting places. Some native plum and a crab-apple tree for their fleeting blossoms in spring, together with a few perennials of the variety one would like for such a composition, completed the garden. This garden brings joy and happiness at all seasons of the year—not so a garden where the plants have to wear winter protection. The birds gather here, and how beautifully they fit into the picture. They, too, like this little bright spot in the forest.

Ancient oaks surround the garden, and two of these sturdy giants loom up where the trail meets the clearing and points the way. A little colony of hazels, situated at the entrance to this home, greet you with their golden catkins in spring as you enter, and at the beginning of the trail to the garden another group of native hazel sings the same song in the April breeze.

This little home has always been inspiring at all seasons. Golden catkins of the hazel against the purple mystery of the woodland, hawthorn snow-white on the clearing, plums and shad in the bottom of the ravines like a mist drifting through these deep valleys, oaks with their golden tassels swaying in the wind against the blue sky of Illinois, fleeting shadows across the clearing and deep shadows in ravine and woodland, the charm of it all in the light of the setting sun and its deep peace when the mystery of night quietly enfolds all—these are its spiritual notes. Each day it has a new message to bring, a new song to sing.

One of our villagers bought an old homestead dating back to pioneer days. Nothing was left to tell the story of those who had lived there except a few apple trees and a few lilac bushes. It faced the rising sun and the waters of Lake Michigan. Along its southern border was a deep ravine. The two outstanding points to consider in planning this home were to preserve this ancient landmark of pioneer occupation and to give it privacy in the greatest measure.

The lilacs were grouped near the house, where they rightfully belonged, and the apple trees of the little

69

orchard were left between the house and the lake. To create harmony between this new home and the reminders of earlier days, I used as a motive in the composition our native crab-apple. It gave right tonal qualities to the whole place, not only during apple-blossom time but also in winter. Beneath the crab-apple trees a carpet of violets were planted to emphasize the festival of the trees. This whole composition represents a friendly home in a friendly garden. It speaks a simple and tender language in keeping with the spirit of the simple life of the pioneer and of a high American culture. No voice of arrogance or dominance disturbs it.

A view of the lake is not to be had until one enters the house. Great, therefore, is the surprise as one looks out upon the changing colors of this beautiful body of water with mountainous cloud effects on the horizon. Only those who have seen this view in spring through a veil of apple blossoms can understand.

An opening was left to the west of the house for a view of the western horizon over tree tops. Here was an ideal place for a few flowers and a pool where birds could find enjoyment. A virgin's-bower trails over the rocky ledge of the pool and is shaded by a graceful crab-apple tree. Visitors are greeted at the door by hollyhocks and lilacs. Trailing against the house and also along the path are prairie roses, inviting when in bloom, but equally inviting in winter when they brighten up the path with their gay colored berries.

Compositions

The ravine was left in its natural state to give the occupants of this home many lessons in what we of today get so little. Along the edge of the ravine the road from the highway to the house curves leisurely. When my memory goes back to this garden home, I vision a lovely composition of plants, sky, and water.

Clients are of all sorts. Those with a real understanding of landscaping are the very, very, few. Some know too much, or have an idea they do, and they are better left alone. Then there are those who want a garden because their neighbor has one, and I am afraid these are in the majority. But there are the few who have a real love for growing things. I have been fortunate in having some of the latter, and just such a person called on me to look over more than a dozen pieces of ground. He asked me for a report on each tract of land so that he might select one of them for his home. These lands differed considerably in character and size.

A quiet place was needed for my friend, as he was very temperamental; therefore, I recommended only two places out of the dozen or more. One of these, which overlooked Lake Michigan, he selected. The land was more or less level except for the lake bluff. Part of it was wooded, but to the west it was open fields.

Peace was my uppermost thought in planning this estate, so the house was placed facing a large peaceful meadow to the west for the brilliancy of sunsets, for shadows over the land at eventide, and for cooling breezes on mid-summer evenings. There was also a view of the lake to the east, but the house was situated far enough away from the bluff to exclude any restlessness or fear of dropping off, which would be most thrilling for the romantic mind, but not so for this man. Toward the lake I planted a group of sugar maple and sumac to reflect the fire of the setting sun backed by the darkness of night.

By cutting away some of the woodland to the west, the meadow was made to flow up to the front door, so to speak. Hawthorns were introduced on the edge of the meadow to give this pastoral scene a feeling of wide open spaces. Here again I used our native crab-apple. It was to put the landscape singing in May, as my client needed the joy that this small tree can give. When it is in bloom, I still slow down on the highway in front of this home to get a glimpse of its friendliness.

To the south of the house was a low depression, usually wet except in extreme dry weather. This suggested a fitting place for a pool and a simple bird bath with water trickling down, under which birds flocked for their daily bath. Back of the pond we placed groups of cedars which loomed up with mysterious forms in the evening light, casting deep shadows over the pool. These cedars were placed in the path of the southern moon, and a plan first formed in the mind, had become an actual fact. Native grasses and narrow-leaved cattails, as well as other native water plants, formed a part of this bit of waterscape, and on the rising land toward the house phlox and other flowering plants sang of the prairies.

Hidden in the cedars and reached by stepping-stones over a narrow place in the pool, we built a circular seat. From here there was an open view toward the lake, which seemed to form into an extension of the depression of which the pool was a part.

A number of years after this composition was completed, my client invited me to dinner, and after dinner we sojourned to the porch, from which there was a view of the pond. As we were watching the shadows of the cedars in the pool on that beautiful moonlight night, my host remarked, "You knew me well."

I asked him what he meant.

"You knew that I was a restless man," he said, "and that I needed quiet and rest after my return from the city. That is why the open expanse to the west. That is why we behold the solemnity and peace of those cedars and the pool tonight. I am indeed grateful. It has meant so much to me."

Along the highway were planted sugar maples. Thirty-six years have passed since then; last fall these maples gave a delightful brilliancy to the highway, greeting the passer-by.

———————

Few people can live where the mind is distracted continually by fear of falling off a cliff, or by sheer daring the elements. The thrill of the rugged and the daring is fine for a short period, but the struggle of life seems to be enough for most of us. To create harmony and friendliness out of a rugged bit of land is at times difficult.

Extreme ruggedness is rather unusual in the locality north of Chicago, but the following place had such in full measure. A very deep ravine skirted the land to the north; Lake Michigan and a short, branching, but also deep, ravine were to the east. The task was to make this bit of ruggedness livable without destroying any of these outstanding contours.

The placing of the house presented difficulties. The site selected was considerably lower than the entrance to the property; so it was out of the question to permit the home to be seen from the entrance, as I have always maintained that the home of man must be in a commanding situation. But having the road follow the large ravine to the north, carving it into the edge of the slope, made a rising grade to the house possible. On entering the property this road presented the picture of a country lane skirted by prairie roses and violets; then it plunged into the depths of the woodland and finally terminated in a large opening at the house.

The land directly in front of the house was low and gathered water which had formed the east ravine during the centuries. Only moisture-loving plants grew in this wet place, and beyond, fires had caused considerable damage to the trees, many of which had died, leaving an open lane. This gave a suggestion for the lawn, which was extended almost to the western border of the property. The outline of this open area was formed to be in harmony with the ravine, its associate. Today one feels that it has been there as long as the ravine and was a part of nature's plan when the deep depressions were carved by the great forces.

It was indeed fortunate that sufficient of the old oak trees were saved from the devastating fires to give proper scale to this landscape; otherwise man would have had to wait more than a century for a complete harmony. Through this clearing stream the rays of the setting sun in late autumn, illuminating the whole landscape. It is most fortunate that the sun sets in this direction at a time when autumn colors are at their height.

The hawthorn is the real motive for this home, and hawthorns were planted at important points on the edge of the meadow to bring out the play of light and shadow over the clearing and to give it breadth and scale. I know of no place where the hawthorn has been able to express itself more beautifully than here. May day is a glorious day on this bit of woodland clearing, and on a May day the owner was laid to rest when the meadow was singing with millions of flowers he could see no more.

The short ravine to the east gave fine access to the lake beach, and this bit of ruggedness was softened by thousands of native flowers planted at the feet of native oaks and maples. Through this ravine a glimpse of the waves upon the beach was possible. The vegetation on the lake bluff was quite dense and was permitted to remain as a protection against the cold winds from across the lake. It was also left to prevent a restless feeling that a steep bluff would give. The land sloped gently from the house toward the northwest, and here a view of the lake and distant shores twenty miles away was permitted.

Ravine landscapes bring outstanding beauty and interest to a home. You never become fully acquainted with the host of friends that find a happy home in these depressions. There are always new friends to greet you. Daily excursions through these ravines bring new discoveries and new delights—a group of lady's-slippers nodding their heads to you, a colony of hepat-icas hanging on a steep slope, a group of bloodroots with their sunny faces close to the trunk of an ancient oak, carpets of anemones or spring beauties, or in midwinter the silvery petals of woodland aster illuminating the deep shadows of dark winter days.

It is far from the prairies of the west to the rocky coast of Maine, to a different landscape with its different beauty—a new world for the prairie mind to understand and to learn to love. The general tone of Maine's landscape is rather dark in comparison with the sunny openness of the prairies. In Maine spruce predominates on the granite bluffs, and granite appears like

76

black loam of the plains. There was much about these hard, rocky precipices that fascinated. Plants strange to me clung to the bold rocks, and beyond was the sea with its changing colors and its vast horizon.

The house, a most pretentious one, hung on a cliff. Level space was very limited, but an intimate and friendly spot on this storm-beaten cliff was most essential. A dry wall of granite, quarried from the adjacent cliffs, produced the garden. This garden had to be a part of the hillside, and folks commonly call it a rock garden, but it is not, so far as rock gardens go. It is a garden of the rocky coast of Maine, and it is a part of that coast.

Spruce, hemlock, and birch formed the background of the garden, and in their deep shadows was built a pool at the foot of a precipitous cliff. Water trickled out of the rocks to keep the pool fresh. On moonlight nights this little garden and pool portrayed a charm amid the ruggedness and the mystery of these ancient cliffs. The garden was carpeted with mint in place of turf. This lovely plant is always in full bloom at the time of year the folks make use of their Maine home.

I have always thought of this place as a folk song with its sweet notes telling the story of the daring flowers that struggle on the cliffs and enjoy the struggle, giving so much joy to others.

Down in the hills of eastern Tennessee, where the dogwood grows profusely and other southern plants thrive, the prairie mind struck another strange world. It was here I first came in contact with southern atmosphere. My client had selected a steep hill, on the top of which he wished to place his home. It was a perfect opportunity for displaying show and pretense. The hill was rugged, and here and there rocks protruded from the earth. When finally the solution of the placing of the dwelling was left in my hands, I was naturally excited and pleased, but the following thought came to mind: "Be careful—you whose childhood feet have never trod the soil you cultivate."

To permit the hill to express what it really was, a high spot in the foothills, and also make it livable, was a real task. Great were the views from this hillside over pastoral valleys and restless hills, on to snow-capped mountains on the horizon, and in placing the home I picked a site where one could see but not be seen.

Lovely views are like lovely pictures or lovely music, something inspiring, something spiritual, something not commonplace and not for every hour of every day. One cannot listen to an opera or a symphony all day long, every day of the week, without finally being irritated. So it was here. To give a feeling of security and restfulness to this steep hillside home was somewhat of a problem. Many of the surrounding views had to be planted out and others softened.

I remember one incident. Loose gravel from the basement had been thrown over the edge of the steep slope, making it steeper in the path of the most important view of the Smokies. A wrong grade at this point would have caused a restless feeling over the whole composition. It was indeed a difficult task to make the foreman understand just what was needed, but after many explanations the work was crowned with success, and the slope at this point was softened by a concave curve. The steep grade was changed into a rolling slope that the eye followed as it gently dropped away into the farming lands below. Today this slope is covered with hundreds of native roses, which preserves its character.

The hills in this section of Tennessee are covered with a great variety of plants, and flowering dogwood grows exuberantly. It was used as the motive and was repeated again and again. Sugar maple and shad, with roses on the slopes, were used to cover up the scars man had made.

Between the house and a distant hill was a deep valley. As I stated before, one problem was to see, but not be seen. Another problem was to make the home livable for man, not for just a month or a year, but for a lifetime. By planting one side of the hill densely, the deep valley was shut out, but something else happened! The eye that caught this planting also caught the distant hill and saw the continuation of the hill flow gradually up to the porch of the home. To me, it was one of the greatest feats I had accomplished in my career.

On the hillside, not seen from the house, we built a swimming hole. It was hewn out of the hillside and reached by intimate trails. On moonlight nights this pool holds a spell over all who visit it. Can you vision moonbeams and deep shadows reflected in the water, or the soft light of the moon lighting up gay flowers planted in the crevices of the rocky slope bordering the pool. With a bit of imagination one can see the dance of wood nymphs.

The rocky background to this swimming hole, as I love calling it, was planted with ferns and other native plants that enjoyed this half-shady situation. Water trickled over the rocks, adding a musical strain besides furnishing water for the pool. At one end the pool disappeared into a level space of iris and other water plants. Above the rocky cliff a council ring was built. From here one could see the river and the river flats beyond, and the fire from this ring sent a friendly greeting to those who lived on the lowlands.

I have always loved this place in the foothills of Tennessee, and I doubt if I have ever enjoyed any of my work more. It was done for appreciative people who understood the real message and the cultural value of these hills amongst which they had placed their home.

~·~·~·~·~·~·~

A home had been built on the gravelly hills of northeastern Michigan. After the house was up, the owner asked me for advice. Again I felt how essential it was to see and not be seen; therefore a grove of maples was suggested for hiding the house, which stood out boldly in the landscape.

Below the house was a level spot ideal for a garden. Like a shrine it hid away in the hillside with a view of the farmlands below. Birches, a note from the north, were here and there in evidence in the woodlands; so birch and shad on a carpet of hepaticas became the garden, and a few colonies of flowers that loved such a shrine, finished it—a small sun opening for rest, a reflection of the delightful quiet nooks so often found in this hilly country. The steep grade of the slope to the garden suggested a pool to attract the birds and for the reflection of crab-apple blossoms that were grouped above the pool.

When autumn haze lies over the land and chilly nights have ripened summer's foliage, the surroundings of this home burst forth into flaming colors illuminating the whole countryside. Slopes covered with fiery sumac literally put the hills in flame, in complete harmony with the sugar maples which crown the hills. Later, when snow covers the land, soft and delicate tints of maple, birch, and shad bring the warmth of the sun's reflection to this hillside home.

For years the message of our great prairies had appealed to me. Every leisure moment found me tramping through unspoiled bits of these vast areas. I wanted to understand their force, their enchantment that called on and on. Then came the opportunity to build a large park on the prairies, at the edge of a great metropolis. No one can realize what such an opportunity meant to me at that time in my life.

The tract of land on which the park was to be built was crossed by an ancient beach dating back to the glacial age. There was nothing very dramatic about this beach. In its earlier days it had been covered with trees and had perhaps been a dry passage for the Indians in the long ago.

This beach gave a fine opportunity for a lagoon at its foot, symbolic of our prairie rivers. It also solved a problem which most park builders have to struggle with—thoroughfares cutting through the park as mother divides the apple pie. In this particular park the most important note in the successful completion of the landscape I wanted to create was the prevention of any thoroughfare cutting across the park from east to west, and this the prairie river did in a most effective manner. One through road, as a communication between city and country, had to be considered. It was gently curved to one side of the tract and caused no inconvenience to motorists. This made the central area of the park quite extensive and also gave enough room between the road and the northern boundary of the park for an intimate natural lane which has become a delightful place for strolling pedestrians and also a place much sought by the birds.

82

Compositions

The interior of the park was made into a large meadow symbolizing a prairie landscape. On the east, the south, and the west boundaries the land was raised with excavations from the prairie river to exclude the city. To the east this elevation formed the bluffs of the river and also a natural division between the playgrounds and swimming pool and the park. The playgrounds and swimming pool were placed adjacent to the city street. They were therefore convenient for those who wished to use them. This proved a very desirable solution for playgrounds in a park devoted mainly to the sylvan beauty that its landscape gives to city folks.

A short road on top of the river bank was built purposely for a view of the setting sun over this prairie landscape. I have always thought, "If the city cannot come to the country, then the country must come to the city." With this in view, both playground and swimming pool were constructed. The pool is surrounded by limestone cliffs covered with wild flowers and backed by heavy planting to give it deep shadows and mystery. The playgrounds represent a natural meadow with pools and woodlands bringing to the city child a bit of Illinois that otherwise would remain strange to him. In one corner, hidden by trees and slightly elevated above the meadow, is a council ring with a fire in the center.

The prairie river has its source in two brooks starting in a bit of broken landscape so harmonious with the river landscape of the plains. Native water plants grace the margin of this river, both for a completion of the water picture and for practical reasons. They prevent erosion by wave action and keep children from falling into deep water. On a promontory along one of the brooks, a player's hill, or stage, for musicals and dramatics finds a suitable place. Across the brook a meadow provides a place for several thousand spectators. From the boathouse, located at the terminal of the river, there is a view of the prairie landscape, of the brook landscape, and a part of the river and its flood plains.

The large meadow, or prairie, as I call it, is surrounded on three sides by heavy planting, as if it were surrounded by a forest. Crab-apple, plum, and hawthorn are predominant on its border, with a forest background of elm, linden, and ash on a carpet of thousands of woodland flowers.

I believe we must know our home environments before we learn about others. We can never become as deeply interested in things foreign to us as in those belonging to our daily life. All the plants used in this park belong to Illinois and are fitting to the soil and climate. These were planted for the youth of our city that they might learn a little more about the great state of which they are a part.

When autumn comes, the glory of the woodlands in their gay coloring is represented here as truthfully and as beautifully as beyond the city, but it is only in this park in a great city of many parks where it is expressed. Often have I heard visitors speak about the beautiful autumn coloring of this park, thinking the plants foreign, not knowing the beauty of their own state. Woodland flowers cover the forested areas, and prairie flowers are permitted to grow unmolested on the edge of a part of the meadow. Here again the outstretched branches of the hawthorn sing in tune with the horizontal lines of the plains.

It is not always granted that a landscaper can see the fruits of his work in near maturity, but such was my good fortune recently when I stood on the bank of the prairie river and enjoyed the peaceful meadow which stretched out before me in the light of the afterglow of the setting sun. My greatest joy, however, was to see that not I alone enjoyed this scene. Others saw the significance of it all, and their silence during nature's great pantomime at the end of the day was the greatest reward I could have received. My early desire to bring to the city dweller a message of the country outside his city walls had become a reality.

But many architects have said to me, "Landscapers are not needed. We do our own landscaping."

I say to them that they do not know what landscaping is. What they attempt in the out of doors is feminine. They have taken an enjoyable task away from the housewife.

One builder asked me what I would do with the surroundings of a home in a hilly country. I answered,

Siftings

"A little back in the valley grow the only birches in this region. Cover the hills with birches. Red cedars also grow here. If you like the showy, plant red cedars for contrast. Then cover the ground with anemones, which were so numerous in this section during pioneer days. Fill the lowlands with plum and hawthorn so that spring in the valley will be festive and sweet with the perfume of plum blossoms. Plant thousands of lilies on the bottom flats and sumac on the hillsides. Along the winding road to the home have thousands of roses greet the visitors. "This," I said, "is landscaping."

Some good soul in the city of Springfield, Illinois, dreamed of the possibility of a Memorial Garden in honor of Abraham Lincoln, and the dream has come true. The garden is simple. It has to be to fit the character of Lincoln. There are large masses of planting to express the greatness of this man and the greatness of the country he served.

The contours of the land suggested lanes or open glades. So there are lanes of redbud and native plum, which are in bloom at the same time; lanes of crab-apple, the most refined small tree gracing the Illinois landscape (and what an inspiration when hundreds of these crab-apples are in bloom in a sunlit lane, backed by the deep and mysterious shadows of woodlands); hawthorn lanes, snow-white in spring and harmonious with the horizontal lines of the plains.

The lower levels, along Lake Springfield, are filled with sun-loving flowers. Here thousands of lilies and phlox greet you in festive colors, reflected in the blue waters of the lake.

Trees cover the hills, trees native to the States in which Lincoln lived and which will grow long and mightily in Illinois soil. Amongst these trees nestle council rings. One of these rings, the largest in the garden, the fire from which can be seen across the lake, has been named "The Lincoln Council Ring." This council ring retreats in a grove of white oak, the sturdiest tree in Illinois, oak that will tell the story of this garden when all the statuary and monuments have crumpled into dust.

Eternal youth is the symphony of this garden. It sings of beauty, of friendship, of honesty, and of strength.

TOWNS

MEN like to live in bee-hives. To think otherwise is contrary to human nature, because man has built bee-hives since the beginning of time. But to say that our present-day cities are ideal is folly. Confusion is evident everywhere.

Such town planning as has been exhibited so far does not interest me. Cities built for a wholesome life, not for profit or speculation, with the living green as an important part of their complex will be the first interest of the future town planner. Most of our large cities have grown under the supervision of the politician or the speculator, and neither has the ability to know what makes a livable city. To guide the city so that it will retain a livable atmosphere is the real purpose of town planning.

It is important that we solve this vital problem because people, like bees, swarm together until they suffocate each other. More space! More elbow room in our cities where buildings are packed together like boxes! Tired workmen as well as children and mothers need the refreshing green of the open areas close to their door. To be worth-while citizens they need the chance to enjoy the soothing rest of starry heavens. Luxurious parks, miles away from their home, do not satisfy.

Much has been written for and against our rural

towns, but we must admit that these towns have a character that is distinctly American. Who has seen any European towns with the over-arching branches of century-old trees shading the streets and the homes of the inhabitants a hundred feet or more away from the street? We have not learned to value the spaciousness of our rural towns, but some day we shall. To import to our cities plans from monarchial countries, with their pompous displays, is a fad reflecting on American intellect. Such plans can never fit us with our love for freedom. I remember motoring through a French town where each home was shut in by high walls, and I also remember that they talked of freedom. Freedom in a prison city! What a joke!

The average American town is one of trees, and let us hope that the openness and the spaciousness of these rural towns will never change. Let us hope that the pavement-minded from our great cities will never be permitted to destroy the true character of these towns, for, say what you will, although the city people like their bee-hives, they have to draw continually from the rural country for fresh thought and renewed vigor. No city can thrive without this continual stream from which to drink.

I have often gazed from the window of the top floor of a hotel in any number of our towns and small cities, and my vision has always been on a sea of green with here and there an aspiring steeple penetrating the mass of foliage. Our larger cities have lost the spirit of these pioneer-built cities. They have become strangers in their homeland. But some day they will revert to the ideals of the pioneer. The city of the tomorrow will demand the living green as a most important part of its composition—the buildings in a garden. Man does well to study nature's way, and if man is to be successful in city planning, it will be man and nature, not just man.

A city with plenty of light and fresh air and the growing green will have no slums. Slums do not thrive under such conditions. When the air of any city becomes so filthy that it will not grow trees, then it is not a fitting place for human beings to live. In the city of tomorrow there will be no smoke to kill the trees or to make life unpleasant or unhealthy for the citizenry. Smoke will not be tolerated where the homes for the thousands are. A more normal life will be the result.

And it will not be at all surprising if the city of tomorrow excludes the automobile. Today the automobile rules, and it destroys the parks as gathering places for the multitudes. Urban travel to the downtown areas of the large metropolis will be taken care of by public conveyance, and the art of walking will come back as a healthy necessity. Why should the motor car be permitted to make life unpleasant for those who are not able to afford such a luxury? It is a poor democracy which allows one crowd to destroy the freedom of another. When any person chooses the con-

gested city for his home, then he must be willing to
live according to definite rules. In that way only çan
so many people live in one small area without trespass-
ing on each other's rights.

I am also certain that there will be no motor-car
parking along the streets in the residential sections,
because sufficient space will be permitted in the city
plan to take care of the motor car inside buildings. A
street lined with automobiles is not a pleasant sight.
The architectural beauty of the buildings and of the
park-like streets is lost in the sea of cars that today line
the streets. The car has done much to destroy the
finer feelings in man, and in the tomorrow it will have
to keep its place.

Efficiency is a much used word, and many, through
no choice of their own, have to suffer for it. Cities
move countrywards in solid formation like an army on
the march destroying all in its path. History or beauty
mean nothing. Let the cities move in groups, leaving
fields and native landscapes between these groups to
penetrate the city and be an harmonious part of the
whole. It has always seemed a great sacrifice when a
fertile field, which has been cultivated for generations,
has to give way to some real-estate project that often
proves unsound and thoughtless. We say it is for efficiency—but what is efficiency?

Our towns and small cities are towns of trees, but
when these municipalities start to grow, the first to
disappear is the tree. The over-crowded city chains the

mind of the people, dwarfing the mind, and a love for the living green, which is a natural heritage, disappears.

Most large centers suffer from a lack of civic consciousness. There is the general complaint that the people cannot be awakened to an intelligent interest in important city issues. The builders of our cities are to blame for this slumber condition. They build a suffocating hive from which the people, or those who can, flee during hot summer months. You will find their shacks creating new slums in our beautiful lake countries. No man can love two homes, and it is neither natural nor sound for a city dweller to divide his interest between city and country. He becomes a useless resident of both. If our cities had plenty of open spaces for fresh air and sunshine, for flowers and trees, they would be a fitting home for their people twelve months of the year, and more time and money would be devoted to the one spot the city dweller calls home.

Mob psychology of the pavement-minded has made our cities the miserable holes they are. In fact, there seems to be little need for the freedom and the beauty of the living green in most of our large centers, for those dwelling within seem to have lost all sense of the beautiful in the city picture. The motor car at the curb and the movie around the corner are more important. Excursions into the country at break-neck speed on highways made by city-minded engineers influence these apartment dwellers slightly. They return home having seen nothing. The human mind is not influenced by excursions, nor by bits of this and a little of that. We are molded into a people by the thing we live with day after day. When the city-dweller becomes contented to live in a desert of brick and mortar, then begins the real danger. When a people becomes mentally ill, it is time to use drastic measures to overcome the degenerating germ.

Chicago was once called a garden city. What has become of the gardens? Great men and great women have grown out of that garden city, but why are there no leaders today? What has happened in Chicago has happened in many other large cities where speculation has been the guiding force.

When a city becomes civic conscious, a town planner is usually asked to make a plan for that city. So far no American city, even with the help of the town planner, has solved the problem of a livable city for all. During my lifetime I have known of only one real town planner, and he was a man of great scope. I asked him if he had complete authority over the district he supervised as town planner. His answer was, "Yes." Then I asked him how he used that authority. He replied,

"My country is the home of a great people. They have contributed to the culture and betterment of the world throughout the ages. It is for me to see that the district over which I have been chosen as town planner will continue to grow great men and great women. If the city council decrees that a woodlot adjoining the city must go so that the land can be used for additional houses, and I feel those woods are more essential for a normal development of that community, then I say those woods must remain, and the city must grow around those trees. Or if a certain field is of more importance to the growing child than if it were filled with buildings, then I say that field must remain."

When he said this, I grasped his hand and answered, "You are the only town planner I have ever met."

If every school building in every village, town, or city throughout our land were surrounded by trees and flowers to represent an oasis in an otherwise barren complex, a pleasing change would take place in the many school districts. Once I suggested that the school board of the city of Chicago procure ten to fifteen acres around every school so that all the schools would lie in parks. Here was an opportunity for young and old to have a healthy community life, to get acquainted with each other, to spend idle hours. It would make the school a center for the life of these different districts, encourage a sound and wholesome city viewpoint. Such areas would appear like wedges from the country beyond, penetrating the dark corners. They would break open the city, letting in sunshine, the perfume of flowers, and the ever-soothing green. I suggested that the park board and the school board unite in this very important task. But my plan was met with defeat because one was Democratic and the other Republican. What has politics to do with the best interests of our children?

Years ago the Chicago Plan was introduced to the

school children of Chicago. I wonder how many of those children know anything about this plan today? How many understand the real purpose of such a plan? How many know there is a plan? Our schools seem so busy teaching youth about other people and about abstract things that there is no time to explain to the child about the community of which he is a part and the part he plays therein. Is it not of grave importance for the people of every city to know the why and the wherefore of their own city? Otherwise, how can any city with so many minds packed into small spaces create anything livable?

A true democracy can never prosper unless the hundreds of thousands born in the congested areas of our large cities can exert the freedom of mind a democracy offers them. Whether they are aware of it or not, they are the voice of the nation, and it should be the concern of every thinking person to see that these people have the mental power and the determination to think for themselves. So far they have failed miserably, and the result is that most of our large cities throughout the land are raging in cunning, trickery, and chaos. The dark nooks and canyons of the large city are the cradle for the tricky and the cunning, and this eventually causes self-destruction. In the city man soon develops mob psychology, and with that his freedom goes. He is no longer a single individual but a tool to be used. His whole view of his community becomes warped, and his interest fades. His mind is absorbed with a conglomeration of things, and his ideals are shattered. Culture and a good life do not grow where gorgeous boulevards border on filthy alleys.

In my many travels I have watched the factories and the factory districts. Here one soon becomes aware that everything is done for the pocketbook of the stockholder, and the worker's interest is seldom given any thought. When the owner and the worker live in the community which surrounds the workshop or the factory, in harmony with each other, where surroundings are fitting to a higher cultural life, then the ideal city community life will exist. There will never be any peace on this earth until we all learn to live and work and think together.

If those in charge of hospitals saw the importance of gardens where the convalescent might while away the hours amongst flowers and growing things, they would do much toward helping a speedy recovery of the patient. One is often amazed at the austere barrenness of hospital surroundings and of the buildings themselves. Hospitals, like schools, have nothing in common with the factory or barrack architecture of today. These buildings ought to be in keeping with their function. The present demand of all for bodily comfort only, with nothing for the spiritual senses, will in the not far distant future be changed.

Can you vision every school, every church, every municipal building in every city with sufficient land for the living green, where the buildings' architecture can be viewed in perspective? As one travels through our large cities, he sees time and again beautiful buildings crowded on all sides by other buildings, and all the city dweller becomes conscious of is the door, which is only one of many doors alongside it.

I like the old courthouse at Springfield, Illinois, surrounded by green and bordered by streets with all sorts of stores. And what an opportunity there is for architectural display on such an open square, where the buildings can be seen in the perspective. On these squares belong all public and semi-public buildings.

I believe the city should own tracts of land for the growing of vegetables and fruits, where the citizens can see and understand that their real existence comes out of Mother Earth, and that the merchant or peddler is only a means of delivery. They should know that they owe their real livelihood to the man of the soil.

And this brings me to the vast areas surrounding our cities. Most of these areas are littered with the corpses of cars, filling stations, advertising signs, hot-dog stands, taverns, and what not. If these areas were turned over to market gardens or for the growing of fruit, one would then enjoy coming to the outskirts of our large cities instead of getting a bad case of nausea at just the thought of the filth and litter one has to penetrate before reaching the city. We little realize what effect such gardens would have on the minds of those dwelling within the city. Just to know that these gardens were there, even if they seldom saw them, would be a benefaction to the city dwellers.

Towns

Large playgrounds for great events should not be forgotten, although it is far more important to develop each individual for a fine body and a fine soul than to create outstanding athletes for the mobs to stand and watch. But it always thrills me when I hear the word "Bowl" for football events. How well it sounds in our language. But the "Bowl" of tomorrow for athletic events of great magnitude has not yet arrived. Stadiums are ugly and unfitting to most towns and cities. Here the man with the tee-square has been permitted to play, and how miserably he has failed. I can see future "Bowls" in a setting of the living green, a part of the city because that is where they belong, and yet something of the country which has penetrated into the city complex, opening it up and giving it light and shadow and fresh air. It has been my good fortune, recently, to work out a community plan for play and recreation. This is the new way. In it lies much hope for the tomorrow.

Cities can be as different as the topography and the climate of which they are a part. Each has the opportunity of building into a beauty all its own. No two hillside cities should be alike. The expression of the land should be carefully considered in the composition. The lake or the ocean-site city can possess a charm that a prairie town lacks. And a prairie town might have the character of which the prairies alone can speak.

I know of one city situated on the low flats bordering a great body of water. Before the city was built, shallow water courses cut through the land, ancient drainages to the sea. How unique and beautiful this city would be today if these depressions had been encouraged to become canals with trees and flowering meadows on either side. Today they are either dumping grounds or have been filled in to make way for efficiency.

Like a tree of the forest, which is a gigantic expression of its associates and friends, so our cities should represent a grand expression of our small towns and villages and the rural country of which they are a part. And the public buildings of these great centers should grow from the beauty of the region and not stand alone as exclamation points.

THE GARDEN

IT WAS evening in the garden. The veil of the approaching night had softened the sharp outlines of flowers and trees. Peace reigned supreme. In the free and unrestrained life of this garden a woodthrush sounded his last note, a reminder of the early evening. Fireflies, like electric flashes from a mysterious world, fleeted about everywhere, becoming more noticeable as the deep shadows of the surrounding woodlands emerged from the darkness of night. This was a time for contemplation and profound thinking. And I asked myself: Would the song of the woodthrush and the beauty of the fireflies have been so fitting and so understanding in a garden of architectural design? Would there be the same harmony in the song and the twinkling light, in the composition of flowers and trees and shrubs, where the freedom to grow was restrained?

On misty mornings the garden has the appearance of floating on water; in a storm, when the heads of the trees and flowers bend in humiliation to the forces, and torrents of rain, like a veil, dim the vivid colors, this garden becomes subdued and in rhythm with the elements. When it is enveloped in a fog which changes the surrounding trees into mysterious forms and colors, the aspect is again changed, but also fitting. These atmospheric changes are all gifts from the Great Master. The garden that does not accept these natural gifts in an understanding way is lacking in its true purpose.

A love for the garden is a vital factor in human civilization. There are records of civilized Indian tribes in Mexico who loved their flowers and trees to the extent that they would fight for their protection. There are songs left to tell the story.

Our country is one complete garden, each section adding its chorus to the great symphony. Sunbeams touching the rosy twigs of gray dogwoods casting their pink shadows on the snow, a cardinal flashing through the air and lighting on the silvery-gray branches of the hawthorn, purple cones of sumac nodding to one another in the northern breeze, the hazel slowly tuning up for its spring song, or a few rocks testifying in a mute way of past ages, are all lovely notes in a garden of winter. All these remind you of the tomorrow that is to come.

During winter the wind comes in crisp and invigorating from across the prairies. At this season of the year the landscape assumes a dreary look to many who do not see and cannot understand. But to others, when the gray arms of the cottonwood are illuminated by the January sun and silhouetted against the blue sky, when sleeping buds are covered with frost sparkling in the winter sun, when the dormant life of millions of flowers is covered with a blanket of snow, when rich plowed fields await the seed that is to feed the millions, and gray and lavender clouds beckon you on over the prairies, the landscape sings a song of rich tonal beauty, a great prelude to dawn, a reminder before the resurrection of life. Our northern gardens should be as fitting in winter as in any other season of the year. A garden with plants that cannot stand the rigid winter weather and have to be protected in all sorts of grotesque ways has lost its poetry and charm.

In a garden that is covered with a white blanket of snow there is a distinct beauty which the far south does not possess. A drift of snow half covering a stone ledge or blanketing a group of prairie roses that send their red berries through the snow, warming the surroundings with the reflected light from the sun, creates a picture of joy and happiness. It is so different from the same garden in summer, yet in both there is the freedom and the urge to be.

When the first snowdrops appear through the snow, singing the song of spring and soon to be joined by the grand chorus of other flowers, when hawthorns are covered with a mantle of white blossoms, when fleeting clouds of perfumed plum blossoms sail through the air, when lilacs remind us of departed friends, then June is on the wing with the full leaf and its cool shade for hot summer days. Too much color during hot summer burns up the soul, but autumn brings back the colors in great profusion. They seem to put warmth into the garden again when it is needed, the chorus of many voices singing their last song before the magic hand of winter puts our friends to sleep and to rest in order to regain their strength and their vigor for another season to come.

Some have said that landscaping is the finest of all arts. Perhaps this is true, but it is waiting to be developed and to assume its right place amongst the fine arts. There can be no question but that the thoughts of the landscaper will be changed in great measure. He will drop the architectural forms and designs; his mind will be free, unshackled by traditions. He must be himself, true to his art. A love for his work must urge him on or there shall never be any art of landscaping worthy of the name. Each environment has a problem to be met, and the landscaper must select in the way of plants, rocks, water, et cetera, those which fit and are in right relation to one another in the whole composition to bring out the character of the landscape in full measure. It is an understanding of the plant itself, of its real beauty, that is most essential in the art of landscaping. Without this there can be no particular feeling, no profound love, nothing to feed the deeper feelings of the human soul. There can be no feeling of love and tolerance and unselfishness.

We shall never produce an art of landscaping worth while until we have learned to love the soil and the beauty of our homeland, and to fit man's accomplishments into its infinite harmony. To separate ourselves from this means degeneration. It is vital, most vital, that we keep this as a prime thought in the art of landscaping. It is a step upward in the evolution of mankind to a higher spiritual goal. Such landscaping is a living expression, where man can commune with the things he loves and the things that have a message from the infinite.

I again repeat: Nature is not to be copied—man cannot copy God's out-of-doors. He can interpret its message in a composition of living tones. The real worth of the landscaper lies in his ability to give to humanity the blessing of nature's spiritual values as they are interpreted in his art. The field is boundless, and there is no need of importing from foreign shores. To the true artist it is like a great adventure into the mysteries of an unknown world.

Nature has already produced variations, as, for instance, in our native plum, our flowering dogwood, and our native crab-apple, and these variations through cultivation may develop more colors. To do this is an honest procedure, and one that will never cause regret. Hybridizing is a different thing, and the results are not always of the best, nor are the varieties produced of lasting quality. I do not believe in hybridizing as carried on by the tricky and dishonest mind which disregards the art of gardening and purposely creates freaks and grotesque plants for profit.

There seems to be a craving for rare things in the minds of many, and we have rare things in nature that are pioneers, that dare travel from the south northward and from the north southward. These indeed are treasures for any garden, providing the garden has been laid out to receive them as fitting plants in the composition.

In the plan of human conduct there is a marked difference between the mind which sees beauty in a simple violet and that which sees it in a pompous rose or dahlia. On the one hand we have a love for the free and untampered flowers of God's creation, and on the other hand for a flower of social ills, sophistication, and conceit.

Until recently I was not acquainted with our common wood lily. By that I mean I had never examined it carefully. This summer I was attracted by the delicate beauty of a few wood-lily blossoms that struggled on a rocky ledge facing the northwest, where wintry blasts send the temperature down to thirty below. These lilies were not pampered nor cultivated nor watered. They had to struggle for their existence. I then thought of some cultivated wood lilies I had purchased from a seed store. On examining these flowers I found them coarse in texture, and in comparing them with the native wood lily I thought they looked overstuffed. I have thought often of this lesson, and I am wondering to what degree the cultivated flowers of our garden dull our finer feelings.

As I look out of the windows into a vast world far to the horizon and far to the heavens above, I can vision the great Father of Waters majestically and powerfully moving toward the blue waters of the Gulf, and beyond that to the shores of many races with their customs and their modes of living, carrying my thoughts into different worlds and different landscapes. I can vision the purple lines of the Rocky Mountains with their heads far into the heavens on the western horizon and the vast snow fields far toward the North Pole, a strange world illuminated by the mysterious rays of the northern lights. This expression of vastness is our country, rich in imagination, rich in freedom of thought, giving freely to the arts of our people.

On a trip to the western coast I had been asked to speak on landscaping to a group of architects at the University of Southern California. I thought of a little grove of sycamores and cottonwoods that had been permitted to remain by someone who had tried to "improve his surroundings." I remembered the composition of these trees in connection with the song of the mocking birds and the laughing waters of a brook that came out of the mountainside. I remembered the soft blue hues of the distant mountains as I saw them through the gray trunks of the sycamores and cottonwoods. It was one of the finest and most intimate pictures that I had come in contact with in that land of the golden sun. I then remembered what a change I saw upon turning about and walking toward the house. Here the improver had set his mark with plants from Japan, China, and Australia, forming a conglomerate mix-up. What a tragedy to a state so rich in native beauty and so different from anything beyond the eastern slope of the Rockies! What a loss to our country! And so to the students I spoke on the soul of Southern California.

Nature talks more finely and more deeply when left alone—left alone as I had witnessed it in this beautiful valley of southern California and had witnessed it so often before. It is here that those deeper sources of life which awaken the spark of creative ability are discovered. Nature never fails in scale. Its work is always in harmony with its environments; therefore the disappointment when man tries to copy it blindly.

As we move onward, the mind of the American in all the arts will turn more and more toward his native land with its unlimited resources and its untold beauty, all leading toward a new and greater cultural life, richer and more wholesome than that brought with us from Europe. American landscaping will grow out of that life which in its various forms makes up the different parts of a finished picture. It is the America which is gradually but definitely moulding us into a people.

It is erroneous to think that tradition has all to do with real progress in civilizing man. Destroy the primitive, let man call on his own creation, and watch the result. When the Greeks thought more about their bodies than about their souls, they disappeared. We boast about great civilizations flowering and dying, and it is a shallow boast indeed. Man is of the infinite, and he can go on forever if he recognizes the Master's work, which will carry man's accomplishments onward and upward to an ultimate greatness and goodness for all.

It seems to me that the rural mind is destined to create beautiful landscapes and gardens, more so than the mind which is confined to the narrow and prison-like space of the city and deprived of the freedom of growth. When I speak of the city dwellers, I do not mean those of town and village who are closely associated with the out-of-doors and spend much time in surrounding meadows and woodlands, especially during childhood days. Education might bring one much, but never the fundamental thoughts that are inherent in those who are associated with the soil from childhood up, and it seems to me that this inheritance is much more important to those who have to do with the living green as the fundamental for their art.

I have had great joy in forming simple compositions with this living green. A study of its song in spring, its coolness in summer, its gayety in autumn, its joys and sighs in winter, with its everchanging lights and shadows has given me many happy days. Its association with our life and its fitness into one harmonious whole have brought to me a lasting satisfaction.

Art must be a guide, a leader, in the evolution of mankind toward a higher spiritual goal. None of the arts is more able to do this than that of the garden. It is a living expression of peace and happiness, and therefore a great influence in the forming of a people. It matters little if the garden disappears with its maker. Its record is not essential to those who follow, because it is for them to solve their own problem, or art will soon decay. Let the garden disappear in the bosom of nature of which it is a part, and although the hand of man is not visible, his spirit remains as long as the plants he planted grow and scatter their seed.

Nature and man's hand must go together. Man has dominion over all life on this earth, but it is not his purpose to destroy that life which God has given him to protect. We are today living in a machine age. What is to follow no one knows, but there is one thing sure: nature will survive. Man in his arrogance and conceit passes away. A bird singing over his grave drops a seed, and out of that seed grows a beautiful tree getting its substance from what was once conceited man. So nature goes on without any vengeance.

Who can realize the supple power and the emotional forces that lie hidden in the misty bloom of the witch-hazel in the purple shadows of the dying day?

AFTERWORD

A Note on Jens Jensen as a Landscape Architect

UNFORTUNATELY, outside the profession of landscape architecture, the name of Jens Jensen is not well known. Who was he? What achievements made him a significant figure in the respective worlds of American landscape architecture and conservation? What is his legacy?

Jens Peter Jensen was born on September 13, 1860, on a farm near Dybbol, Slesvig, Denmark, into a prosperous farm family. In 1864 Germany had successfully invaded and taken possession of that part of Denmark, and hence Jensen grew up under German rule. Nevertheless, he remained passionately Danish at heart throughout his life, an attitude that was fostered by his family when he was sent to a Danish folk high school at Vinding, Jutland, and then to agricultural college at Tune, in the Danish province of Zealand. This strong sense of Danish culture and love of the Danish landscape, Robert Grese has noted, served as a focal point in Jensen's life, and he returned often to his boyhood memories to convey a sense of place, both in his writings and in his landscape designs: "When the first flow-

111

ers appeared in spring, father made pilgrimages with his boys to the bluffs towering above the open sea. Can anyone realize what it meant to those who had been shut in for months to be greeted again by the warmth-bringing rays of the sun and the lovely green it brought forth, changing the earth into a new beauty?" (*Siftings*, p. 14).

Between 1880 and 1883, Jensen served in the German army and was assigned to a regiment of the imperial guard in Berlin. This experience intensified his lifelong dislike of militarism and autocratic rule. In 1884, after completing his military service, instead of returning home to take over the family farm, as his family intended, he bought passage to America for himself and his sweetheart, Anna Marie Hansen, and married her upon their arrival in the United States.

Jensen worked briefly as a laborer in Florida and Iowa before moving to Chicago in 1886, where he took a job as a gardener with the West Chicago Park District. His performance in that position led to a series of rapid promotions within Chicago's park system, first as foreman at Union Park and then, in 1894, as superintendent at Humboldt Park. An abrupt dismissal in 1900 for political reasons allowed him to begin his career as a private landscape architect in earnest, and he was soon designing the estates of wealthy Chicago industrialists on the fashionable North Shore. As a result of Jensen's association with this group of influential

Chicagoans, Stephen Christy notes, Jensen's reputation was restored and he was rehired in 1906 by the West Park Commission as general superintendent and landscape architect, a position he held until 1920.

During these years in Chicago, with major funding programs in place, he planned landscape improvements in the existing West Parks and coordinated the acquisition and design of many other parks. For this he is credited with introducing the "Prairie Style" into the landscape architectural profession. Some of Jensen's greatest public work was implemented during this period, most notably Humboldt Park and Columbus Park, because it was in these and other parks that Jensen produced a landscape design that was modeled on the composition and structure of the regional landscape. As he himself writes, he wanted "to give the people of Chicago a bit of native Illinois, something most of them have little chance to see or feel. So the prairie river which follows an ancient beach was planned with the prairie landscape beyond, reflecting the evening light as it is seen from the bluffs of our prairie rivers."

While Jensen was on the Chicago payroll he still maintained a limited private practice, but in 1920, with funding and support for his park work in Chicago on the decline, he left the city's employment for good to devote full time to his landscape practice, which he operated out of a studio in Ravinia, Illinois. The design work covered the gamut—from small residential properties to Fair Lane, Henry Ford's magnificent estate in Dearborn, Michigan, with its Great Meadow. It also included landscape designs for some of the residences that were created by Prairie School architects Louis Sullivan and Frank Lloyd Wright. Chicago's most rich and famous—Armour, Cudahy, Florsheim, Kuppenheimer, Rosenwald, Ryerson, and Swift—all turned to Jensen for their private landscape work, as did other clients throughout the nation.

The Ravinia office practice continued from 1920 until 1934, when his wife, Anna Marie, died. The following year he turned his practice over to his son-in-law, Marshall Johnson, so that he could move to his 128-acre parcel of land that he had assembled on the western tip of the Door County peninsula near Ellison Bay, Wisconsin. There, with the help of his secretary, Mertha Fulkerson, who had worked in the Ravinia office for the previous decade, Jensen would realize his dream of establishing a school where people from all walks of life could be in close touch with nature and learn "lessons of the soil." Significantly, Jensen himself said that, in naming "The Clearing," he was referring not to an open field in these beautiful north woods but to the clearing of the mind that could occur there.

Jensen spent the last sixteen years of his life at The Clearing, from 1935 until 1951, and it was there that he wrote *Siftings,* his major literary work. During these years, his students—often young men from the city who aspired to become landscape architects—would come north to this splendid retreat and spend several weeks or even months listening to the master Jensen, and "learn by doing" in the gardens, fields, and forests of The Clearing. Mertha Fulkerson, in *The Story of The Clearing* (1972), which she wrote with Ada Corson, recalls one of Jensen's most quotable prescriptions for teaching students his own subject:

"First grow cabbages. After that, plant a flower. When you have successfully grown a flower, then you can start to think about growing a tree. After watching a tree grow for several years, observing how its character develops from year to year, then you can begin to think of a composition of living plants—a composition of life itself. Then you will know what landscape architecture is" (p. 29). Jensen, no doubt, was only half-serious when he uttered such a thought, but he was fully serious about the need to understand the relationship between plants and their use in a design: "Every plant has fitness and must be placed in its proper surroundings so as to bring out its full beauty. Therein lies the art of landscaping" (*Siftings*, p. 41).

After Jensen's death, the pattern of classes that prevails even today—one-week sessions on a wide variety of subjects in the arts, humanities, and natural history— was established by Fulkerson as a way of keeping The Clearing open and making its truly calming environmental effect accessible to about twenty-five people each week, from June through October. In keeping with his dislike of the military mind and with his love of the democratic ideal, Jensen had planned that, amidst the setting of nature's gifts (the beautiful north woods, Lake Michigan's historic Green Bay, and the still clean air), the classroom and subjects would extend far beyond walls and curricula definitions.

It should also be noted that, from his earliest days in the Chicago region, Jensen delighted in taking trips to the countryside, and he spent weekends on such journeys with his family, friends, and members of the pioneering organization, "Friends of Our Native Landscape," which he established in 1913 to promote conservation and preservation of America's remaining natural landscapes. Among his acquaintances was Henry Cowles, a well-known botanist who was especially interested in ecological relationships. It is likely that Jensen learned from Cowles not only the identity of many of the native plants of the region, but also much about the ecological concepts of succession and plant communities that became a trademark of a Jens Jensen landscape design.

Jensen, in addition to his work in landscape architecture, was one of the nation's preeminent conservationists. Long before other activists took over the idea, for example, Jensen foresaw the need to preserve good examples of the dunes, forests, prairies, and wetlands that were native to the Upper Midwest. He personally surveyed the lands along the Des Plaines, Sac, and Calumet rivers, and developed plans for the Cook County (Illinois) Forest Preserves. He was instrumental (often with the help of Friends of Our Native Landscape, among whose members was Stephen Mather, the first director of the National Park Service) in preserving other areas, such as the State Park System of Illinois, the Indiana Dunes (now a state park and a national lakeshore as part of the National Park System), eight county parks in Door County, Wisconsin (which, thanks to the publicity derived from a *National Geographic* article published twenty years ago, has become known as the "Cape Cod of the Middle West"), and the spectacular Ridges Sanctuary in Bailey's Harbor, Wisconsin.

Today Jensen is being rediscovered for the genius he truly was, and many followers (or fans, as Charles Little calls them) believe that Jens Jensen will eventually take his rightful place alongside the other landscape greats—Andrew Jackson Downing, Frederick Law Olmsted, H.W.S. Cleveland, John Nolen, and the like. Jensen was a master at understanding the ecology of a regional landscape, and he fully embraced the concept of using native species almost exclusively in his landscape designs. Yet he did so with broader intentions in mind: "To try to force plants to grow in soil or climate unfitted for them and against nature's methods will sooner or later spell ruin. Besides, such a method tends to make the world commonplace and to destroy the ability to unfold an interesting and beautiful landscape out of home environments" (p. 42).

This perception of design led to Jensen's extraordinary manipulation of outdoor space and his ultimate belief that no designer could claim to be able to copy or reproduce nature exactly. It was likely for this reason that Jensen never produced a landscape that included a true prairie restoration or re-creation. His reasoning is clear, and it may best be appreciated in the following passage: "I again repeat: Nature is not to be copied—man cannot copy God's out-of-doors. He can interpret its message in a composition of living tones. The real worth of the landscaper lies in his ability to give to humanity the blessing of nature's spiritual values as they are interpreted in his art. The field is boundless, and there is no need of importing from foreign shores. To the true artist it is like a great adventure into the mysteries of an unknown world" (p. 105).

Jens Peter Jensen died on October 1, 1951, at the age of 91, in his upstairs bedroom at The Clearing. His ashes are buried in Memorial Park Cemetery, near Skokie, Illinois. At the time of his death, the *New York Times* described him as the "dean of American landscape architecture." Perhaps as more Americans come to know his writings and his landscape designs, they will better appreciate why to many fans Jens Jensen will always be a legend.

DARREL G. MORRISON

CHARLES E. LITTLE, the series editor for American Land Classics, has written six books on landscape and conservation. He is the books columnist for *Wilderness* magazine and contributes articles and reviews to other periodicals as well. He is the editor of *Louis Bromfield at Malabar* (1988), also available from Johns Hopkins, and has just completed a seventh book, *Greenways for America,* scheduled for publication in 1991 by Johns Hopkins. He lives in Kensington, Maryland.

DARREL G. MORRISON is dean of the School of Environmental Design at the University of Georgia and a fellow of the American Society of Landscape Architects. He is a nationally prominent authority on the restoration of prairies and other natural landscapes, and from 1981 to 1988 was coeditor of the award-winning periodical *Landscape Journal.* He lives on the Oconee River, near Watkinsville, Georgia.